Western Poultry Book

The Chicken Business From First To Last – What To Do and How To Do It

by Mrs. A. Basley

with an introduction by Jackson Chambers

The World's Largest Selection of Vintage Poultry Books

www.VintagePoultry.com

Introduction

I am pleased to present yet another title on Poultry.

The work is in the Public Domain and is re-printed here in accordance with Federal Laws.

As with all reprinted books of this age that are intended to perfectly reproduce the original edition, considerable pains and effort had to be undertaken to correct fading and sometimes outright damage to existing proofs of this title. At times, this task is quite monumental, requiring an almost total "rebuilding" of some pages from digital proofs of multiple copies. Despite this, imperfections still sometimes exist in the final proof and may detract from the visual appearance of the text.

I hope you enjoy reading this book as much as I enjoyed making it available to readers again.

Jackson Chambers

Self Reliance Books

Get more historic titles on animal and stock breeding, gardening and old fashioned skills by visiting us at:

http://selfreliancebooks.blogspot.com/

INTRODUCTION

In the hope of helping beginners and others of my friends in the poultry business, and in response to urgent requests for a book on poultry culture from my pen, I wrote a small volume three years ago. The whole edition was sold in a year, and on account of the interest taken in it and the demand for something more, a second edition was issued. This edition is now exhausted and a third edition is now offered, with additional chapters and up-to-date information on breeding, fireless brooders, and other new features in the poultry industry.

The book is a synopsis of many chapters of my "Woman's Work in the Poultry Yard" and other talks on poultry, and embodies the personal, practical experiences I have been through myself in many years of pleasant work in the poultry yard. Its object is not necessarily to urge anyone into the business, but to encourage and help beginners and especially newcomers, not back East but on the great Pacific Coast and in the Western States, where conditions differ materially from those in the East and where there is an increasingly large demand for both poultry and eggs; where the poultry business is about as profitable as any that can be undertaken and a good living may be made in the pure air and sunshine by any industrious man or woman.

Having for many years been lecturer at the Farmers' Institutes in the Extension Courses of the University of California, for four years instructor in poultry husbandry at the poultry school of the University of California, and having been editor or associate editor of four agricultural magazines and several other newspapers on the Pacific Coast, many questions have during this time been propounded to me relating to the poultry business, its difficulties, the troubles of poultry raisers and the ailments of fowls. Some of these questions will be found in this book with the answers to them, also remedies for the diseases or ills of fowls in this climate.

Hoping and feeling sure that my little book, which is the only book dealing with the climatic and other conditions "Out West," may prove a help to all its readers, I am,

Very cordially your friend,

Mrs. A. Basley

MRS. A. BASLEY

TABLE OF CONTENTS

CLASSIFIED INDEX

MRS. BASLEY'S CONTINUOUS FRESH AIR HOUSE AND SCRATCHING SHED.

PART I.

COMMON SENSE POULTRY HOUSES

The poultry business is one of the most fascinating as well as the most profitable, considering the amount of capital invested, in the West. The conditions here, however, differ so greatly to those in the East and other localities, that the ways of treating the fowls must also be different. The needs of fowls do not vary; the resources of the places do, and the success of the poultry raiser greatly depends upon adapting the conditions of the locality to the need of the fowls.

Nothing is more important than the proper housing of chickens. The style of house a man builds for his birds will depend upon his means and inclinations. It is not always the most expensive house that gives the most eggs. In planning poultry houses and yards, two or three principles should be firmly held in mind: First, the house must have a liberal supply of oxygen, which can only be supplied by perfect ventilation; secondly, it must be free from draughts and be dry; and, thirdly, be easily accessible to the attendant, not only for cleaning and spraying, but to enable one to handle the fowls when on the perches. It should also be large enough to avoid crowding of the fowls.

The laying hens should be kept in yards in permanent houses, easy of access, whilst the young and growing fowls will do best on free range with movable houses, called sometimes colony houses. These give the best results.

After many years of experience here, the writer has found that there are two classes of houses admirably adapted to the needs of the fowls and to this climate. These are called the open front or the "fresh air" house and the "mushroom" house. What is meant by an open front house, is a house enclosed on three sides and roof, with one side open to the fresh air. This style house can be constructed as a separate and movable house or as a continuous and

W. G. Suit's Double Open-front Houses and Scratching Shed, Bandini Ranch.

scratching shed house. A plain open front house without a scratching shed attached, is used in many places as a colony house where fowls have free range or where they are kept in an orchard.

The "mushroom" house is built tight on four sides and roof, without any floor, and is raised from the ground about twelve inches.

Cuts of both of these styles of houses will serve to show their construction.

A "fresh air" house that proved excellent and which I used for years on my ranch, was one hundred and twenty feet long and ten feet wide. It was divided into six houses with scratching pens. I also had another which suited me well. It was eight feet wide and a hundred feet long; besides that, I had twenty colony houses for the young and growing stock, and two brooder houses.

The continuous house and scratching shed of which I give a photograph and part of ground plan were built of flooring, tongued and grooved.

The other house was of boards, battened, and the colony houses of resawed redwood or of shakes. Some were of rubberoid or building paper.

Many of the artistic looking house plans which may be found in poultry books were planned by men who never owned a chicken, and if built in this, or in any other climate, would be highly unsatisfactory. The plans here described have all been used either by myself or by successful poultry raisers. I have seen them all and can assuredly recommend them for use on the Pacific Coast.

CONTINUOUS HOUSE AND SCRATCHING SHED.

The houses I am describing are of the inexpensive kind, for so great is the variety of plans of houses designed for fowls that it would be impossible to mention them all in a short article. We will,

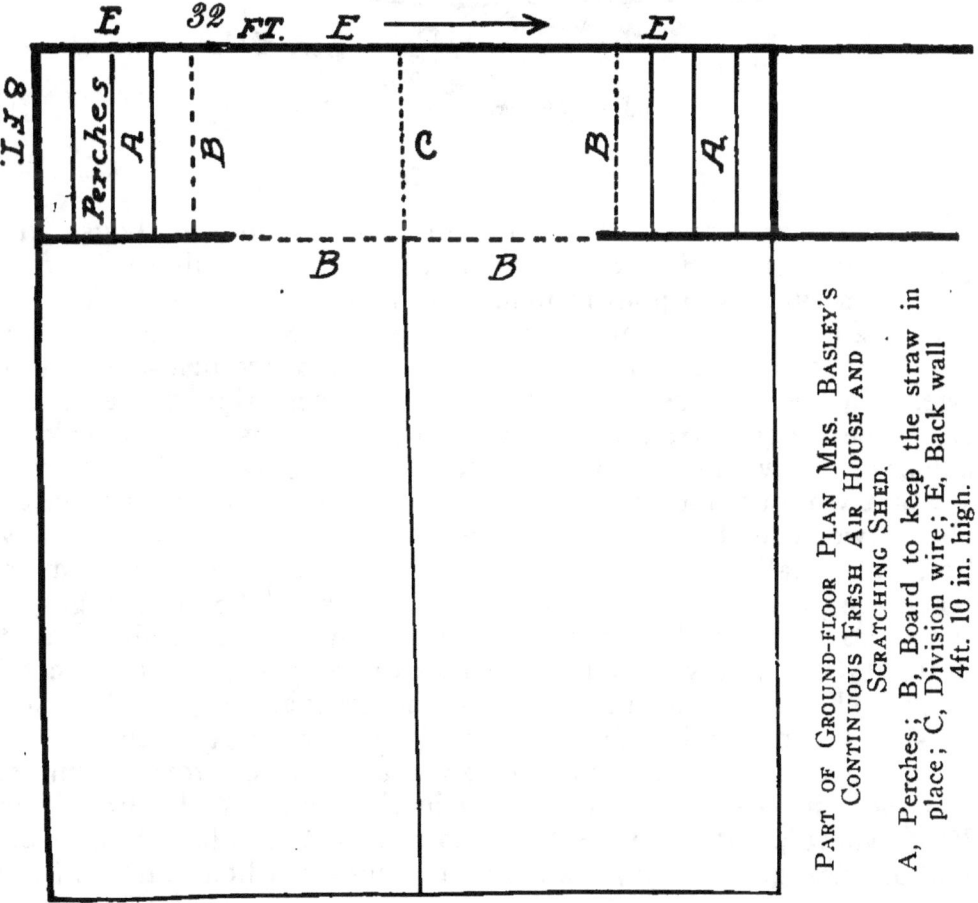

PART OF GROUND-FLOOR PLAN MRS. BASLEY'S CONTINUOUS FRESH AIR HOUSE AND SCRATCHING SHED.

A, Perches; B, Board to keep the straw in place; C, Division wire; E, Back wall 4ft. 10 in. high.

therefore, consider only a few of the cheapest and most satisfactory small houses adapted to this climate.

The first requisite in the house is pure air. To secure this the ventilation must be at the bottom. Some people think that the bad air ascends, but this has been proved a mistake—the foul gases descend; the pure air and the warm air are lighter, and they rise and we want to keep them in, but if we have an opening for ventilation at the top or near the top of the house, we lose the warmth. A loss of warmth at night in the winter means a loss of eggs, or more food is needed to supply this loss. The ventilation should either be at the bottom, or one entire side of the house should be left open.

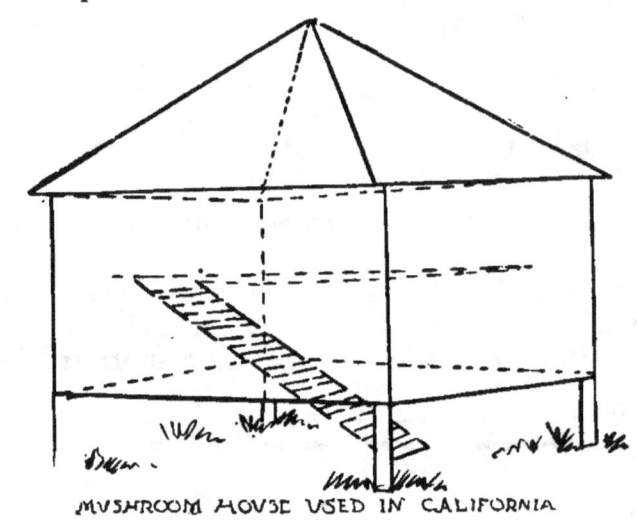

MUSHROOM HOUSE USED IN CALIFORNIA

A Variety of Houses

The accompanying rough little cut of a "mushroom" house will give some idea of the bottom ventilation. Houses like this were used by a successful poultryman. He made a light frame five feet square and five feet high. This he covered with canvas and the roof he made of rubberoid roofing. He left a space below of ten or twelve inches. These "mushroom" houses were tipped over every day to be sunned or cleaned. I improved upon his plan by making a door of one whole side, for I wanted to be able to handle my fowls at night without tipping the house over. Perches should be placed about twelve inches above the open space, and in case of heavy breeds, a small ladder or run board should be placed for them to reach the perches easily when going to roost. The advantages of such a house are its lightness and the free circulation of air without draughts on the fowls. These houses can be covered with matched lumber, shakes, canvas, burlap, rubberoid, or even common domestic muslin, which may be oiled or painted with crude petroleum.

The open front house is admirably adapted to California climate. It is now meeting with favor even in the rigorous climate of the East, where poultry raisers begin to realize the value of fresh air without draughts, if they want to have vigorous hens that will lay

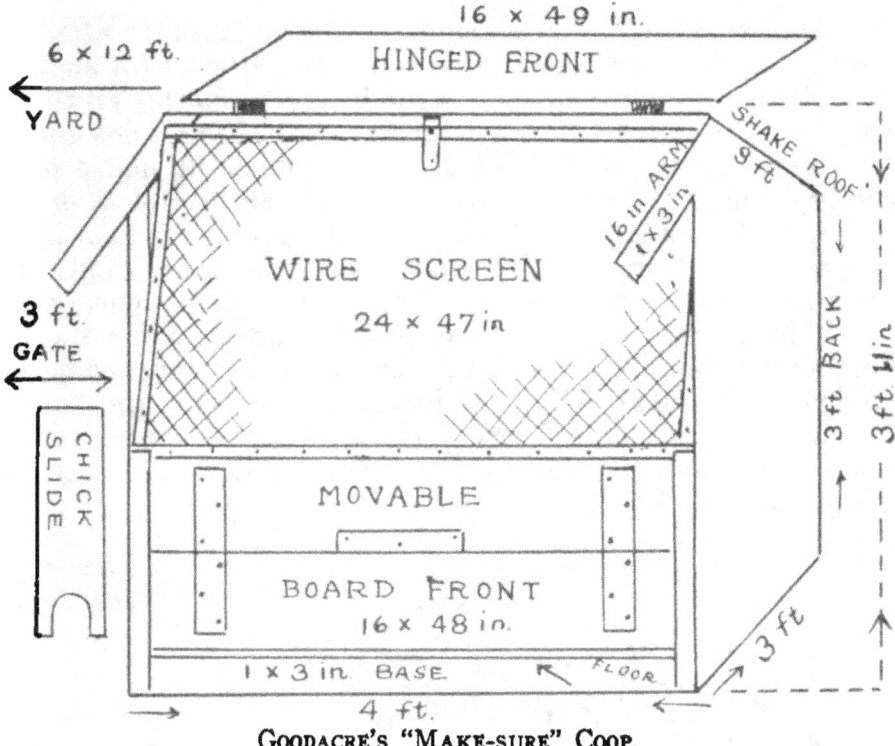

GOODACRE'S "MAKE-SURE" COOP.

eggs in the winter time. I have been using the open front houses of various sizes for over twelve years, and can assert that they are the only kind I ever want to use. Another style open front house that I have seen and like very much is fifteen feet by eleven feet six inches, and is seven feet high at the back and four feet at the open front. It is constructed of rubberoid or malthoid and is almost vermin proof. It is divided in the middle by chicken wire, so forming either one house or two as required. The roof is first covered

GOODACRE'S COLONY COOP.

with two-inch chicken wire to support the rubberoid. At the bottom of the walls, next to the ground, it is boarded up for about two feet all the way round; this is to keep in the straw, for all the floor space of the house is used as a scratching pen. The sides and back above these boards are made of panels of rubberoid, nailed to light frames without the chicken wire. These panels are taken down on all fine days to sun and air the house. The panels are kept in place by large wooden buttons. The front is entirely open or only closed by chicken wire, except when it rains, then a burlap curtain is let down. The perches are near the back of the house, about six inches above the dropping boards. The dropping boards are made of the rubberoid on frames. They are four feet wide and are placed on cleats two feet from the floor. This is a double house and each side will hold from twelve to twenty hens. The above description is of the Hoffman house.

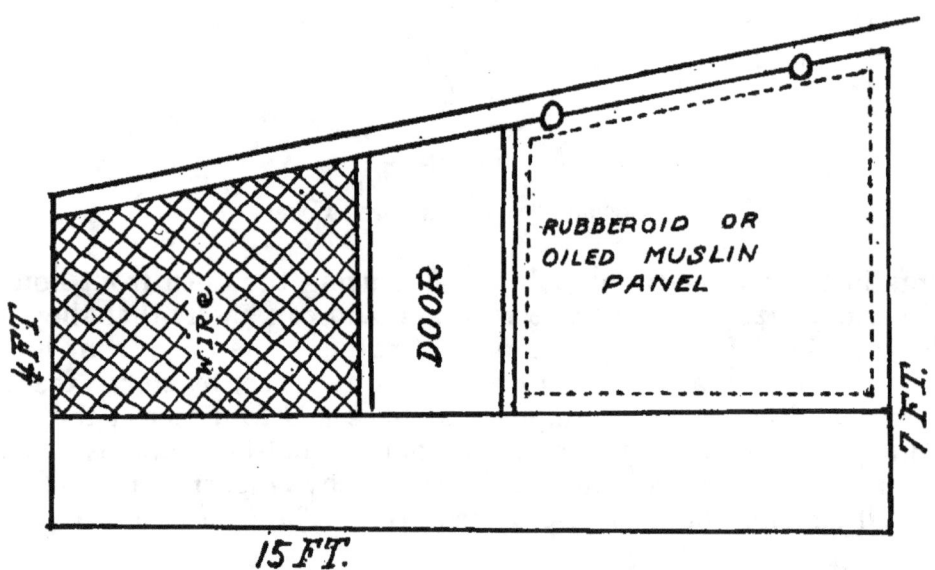

HOFFMAN'S COMBINATION OPEN-FRONT HOUSE AND SCRATCHING PEN

A cheap and substantial house can be made of two piano boxes. The simplest way to make such a house is as follows: Removing the backs of the piano cases, place the cases back to back thirty inches apart, on light sills. Use the boards which were the backs to fill up the thirty inches on the sides and roof; cover the roof with rubberoid or with oilcloth, and you have a comfortable house, that will hold about a dozen or twenty hens, at a small cost. The front of the piano box house should either be hinged so it can always be kept open, except during the rain, or it may be entirely dispensed with and a burlap curtain used to keep out the rain. The cost of this piano box house is about three dollars.

Inexpensive Colony Houses

An inexpensive colony house is pictured below. This house is of resawed redwood, four by six feet. It is light and easily moved.

The front is on hinges and it is always kept open except during rain, and when it is closed it only comes down six inches below the perches, leaving an open space of about fifteen inches across the entire front.

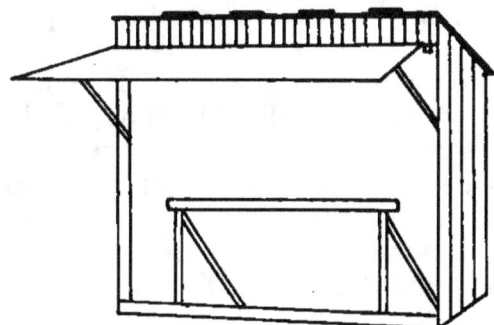

OPEN-FRONT COLONY HOUSE WITHOUT SCRATCHING SHED.

Still another style of colony house and one well adapted for use in an orchard or in the colony plan has been in use for some years on a large poultry ranch in California. The house is eight by ten feet and two feet to the eaves; all the framework, including the runners, is of two by three-inch stuff, and the walls and ends are of one by twelve-inch boards, shiplapped so as to avoid using battens. The rafters are five feet four inches long, and three pairs are used; a one by six inch strip is run all around the outside of the roof to form the eaves and also to make it tight; eight pieces of one by four are used for sheathing, and the sawed shakes are close, so that there is no draught from that source; the only opening is from the front, which is open at all times. The houses do not require cleaning, for they are on runners, and are slid along about fifteen feet each time. Thus they are on fresh ground and much cleaner than one could do it in any other manner.

GOODACRE'S "MAKE-SURE" COOP IN ACTION.

The Two-Story House

Among the hen houses, or chicken coops, as some people prefer to call them, that are being used very satisfactorily west of the Rockies, must be mentioned the two-story houses. These are especially adapted to the "intensive" method of poultry culture, and for limited space.

In conclusion, to quote Mr. Harker, "If every poultry keeper on the Pacific Coast would make his roosting houses absolutely draught proof on three sides, yet leaving the front entirely open so that the fowls have an abundance of pure air, yet not to be exposed to a draught, the manufacturers of roup remedies would have to go out of business, for this disease would then be comparatively unknown from Seattle to San Diego."

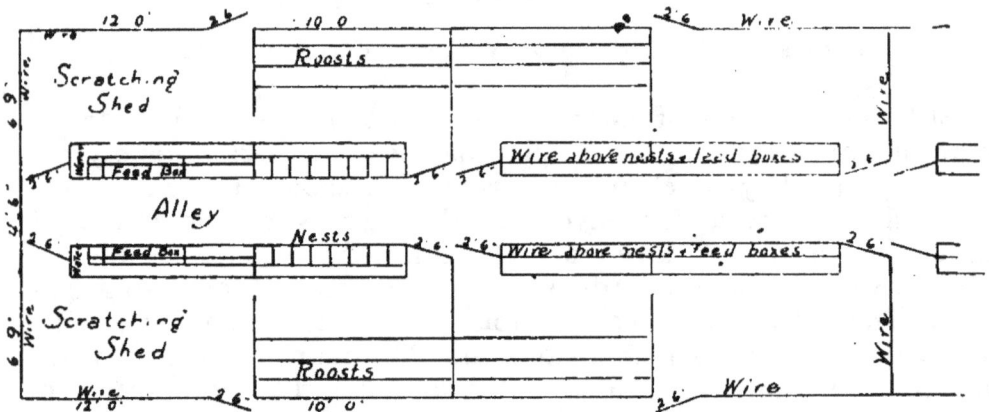

W. G. Suits' Ground Plan of Double Houses, Bandini Ranch, Riverside, Cal.

Painting the Houses

For painting the houses I have found nothing better than the crude petroleum. I add to it for all my houses, red Venetian paint, mixed with a little kerosene or distillate oil to thin it. This colors them a handsome chocolate. Creosote stain of a dark green is also a very good color, harmonizing well with the landscape, and both of these are preventive of mites and keep their color well for several years. A good whitewash also is quite suitable.

A cheap green stain is made of six pounds of yellow ochre mixed with one gallon of kerosene, adding lamp black until it is of the desired shade of green. I think that mixing the yellow ochre with the crude petroleum which you can get at the oil wells, without using the lampblack, would make a very desirable green, but I have not tried it. Another good green can be made by mixing chrome green (dry) with one gallon of linseed oil, four gallons kerosene, and one gallon of water. The color is a matter of taste after all, and I am only describing the inexpensive methods successfully used.

Here is a recipe for whitewash which is unrivaled. It will stand the wear and tear of the elements for a long time. Anyone by adopting the following formula cannot help attaining success:

Into a tight box or barrel put five or six gallons of hot water in which has been dissolved four or five pounds of coarse ground salt. Into this put a pail full of the best lime obtainable. The large lumps should be broken into quite small pieces. Immediately cover the barrel and cover with a heavy weight, in order to keep it in place when the lime is slaking, for the uplifting power of the boiling mass will be surprisingly great. After a few moments uncover and stir the mixture to the bottom with a long stick, then recover and keep closed for a day or two. When fully slaked the lime should be of the consistency of thick cream. When applied to hen houses or a fence, it should be thinned with water to the consistency of common paint.

If too much water is used in slaking, the lime will be drowned and as a result the wash will be thin and watery. If not enough water is used, the lime will "burn" and granulate. If properly slaked, the mass will be smooth and free from lumps.

When applying the whitewash, dip out a sufficient quantity into a pail, then stir in a handful of cement. This will cause the wash to firmly adhere to the surface to which it is applied. It will be a dazzling whiteness and will "lay on" like paint.

An excellent plan when whitewash is to be used about the hen house, chicken coops, etc., is to put in a liberal quantity of crude carbolic acid.

This may be a lengthy description of the simple process of making whitewash, but anyone will find the recipe first class. The old-time method of slaking lime in cold water and applying the weak solution is very unsatisfactory.

ONE HATCHER OF THE GARDENA HATCHERY, CONTAINING FOUR THOUSAND EGGS.
THERE ARE TWENTY-THREE OF THESE IN THE BUILDING, MAKING THIS THE
LARGEST HATCHERY IN LOS ANGELES COUNTY.

WHAT VARIETY TO CHOOSE

"Poultry for profit" is the slogan. We are all looking more or less for the "almighty dollar." Every week, almost every day, I am appealed to for information as to which breed is the most profitable. I can and often do tell which breed I have found the most profitable in the twenty years I have bred, but I cannot decide for another person what his or her likes or dislikes may be, nor can I tell what poultry will suit another's location or market. That, each one must decide for himself or herself, and then get the best of that breed to start with.

A hint as to what to start with may help some of our readers. First of all, study your market, decide whether it requires a brown or a white egg, and choose accordingly; secondly, decide what you will do with the surplus chickens, although this may seem like counting the chickens before they are hatched. Will you sell them as broilers and fryers or use them as roasters or capons? Thirdly, it is always a good plan to look ahead and choose a breed with a prospective value and demand—one of the breeds that may be rare in your neighborhood, or one of the newer breeds, such as the Orpingtons, Campines, Faverolles or Anconas. Choose a breed for which there is likely to be a large demand for eggs for hatching and for breeding stock. Or else take one of the best old breeds that you know will make you money from the start. Whatever breed you decide upon, get the best of that breed, and from a reliable breeder.

Different Breeds

A brief review of the different classes and breeds of domestic fowls may be of use to beginners. There are a large number of breeds in this country suitable to any branch of the business, with all colors of plumage and size. Some especially adapted to the farm, others to closer confinement, as on the city lots, and still others—like the beautiful little bantams—adapted to lawns and front yards.

The American Class

The American class consists of what are called the dual-purpose fowl. That is, they are good for market as well as excellent layers, so when their day of usefulness in the egg basket is over, they can end their existence on the table. This class gives us the Barred, Buff, White, the Partridge, Silver Pencilled and Columbian Plymouth Rocks, the Silver, Golden, White, Buff, Silver Pencilled, Black and Columbian Wyandottes, the Single and Rose Comb Rhode Island Reds, the Buckeyes, the Black, White and Mottled

BLUE RIBBON RHODE ISLAND RED PULLET.

Javas, and the American Dominique. Of the list no doubt the Barred Plymouth Rock is the best known and most popular; it may be said to lead the American class. Next to it in popularity is the White Plymouth Rock. This breed led in numbers at a late show in Madison Square Garden in New York, which is a sure indication of its popularity. The order of the rest might be given as follows: White Wyandotte, Rhode Island Reds, Buff Wyandotte, Buff Plymouth Rock, Silver Wyandotte, Partridge Wyandotte, Golden Wyandotte, Buckeyes, American Dominique, Black Java.

The standard weights of the above are as follows: All of the Plymouth Rocks, cock, 9½ pounds; cockerel, 8 pounds; hens, 7½ pounds, and pullets, 6½ pounds. All of the Wyandottes, cock, 8½ pounds; cockerel, 7½ pounds; hen, 6½ pounds; and pullet, 5½ pounds. The Rhode Island Reds, cock, 8½ pounds; cockerel, 7½

WHITE WYANDOTTES PRIZE WINNING PEN OF A. L. JENKINS, SEBASTOPOL, CAL.

pounds; hen, 6½ pounds; pullet, 5 pounds. Buckeyes, half a pound heavier, except pullets. The Javas are of the same weight as the Plymouth Rocks, and the American Dominiques, cock, 8 pounds; cockerel, 7 pounds; hen, 6 pounds; pullet, 5 pounds.

BARRRED PLYMOUTH ROCK.

The Mediterranean Class

In the Mediterranean class we have the Single and Rose Comb

Brown, Single and Rose Comb White, Black, Buff and Silver Duckwing Leghorns; the Black and White Minorcas; the Blue Andalusians, the Black Spanish and Mottled Anconas.

The Mediterranean class is particularly well adapted to the climate of California, which greatly resembles that of their home in the old countries.

In point of popularity and merit, the kinds might be classed as follows: White Leghorn, Brown Leghorn, Black Minorca, Blue Andalusian, Black Spanish, Rose Comb Brown Leghorn, Rose Comb White Leghorn, Buff Leghorn, White Minorca, Anconas, Silver Duckwing Leghorn and Black Leghorn. The Black Minorca, White Leghorn and Black Spanish give the largest sized eggs.

All of the Mediterraneans have white shelled eggs. There is no standard weight to the Leghorns. They are small birds, weighing 3 to 5 pounds. Of the Black Minorcas the cock weighs 9 pounds; cockerel, 7½ pounds; hen, 7½ pounds; pullet, 6½ pounds. The weight of the Andalusians are, cock, 6 pounds; cockerel, 5 pounds; hen, 5 pounds; pullets, 4 pounds.

The Black Spanish weights are, cock, 8 pounds; cockerel, 6½ pounds; hens, 6½ pounds; pullets, 5½ pounds. These lay an extra large, handsome, white-shelled egg.

The Blue Andalusian has the unique distinction of wearing the national colors—red, white and blue—its plumage being blue, its face and eyes red and its ear-lobes white.

The Asiatic Class

The Asiatic class consists of the Light and Dark Brahmas, White and Black Langshans, the Buff, Partridge, White and Black Cochins. In point of popularity, they would be about in this order: Light Brahmas, Black Langshans, Buff Cochins, Partridge Cochins, Dark Brahmas, White Cochins, White Langshans and Black Cochins. The standard weights are: Light Brahmas, cock 12 pounds, cockerel 10 pounds, hen 9½ pounds, pullet 8 pounds.

Weights for Dark Brahmas are: Cock 11 pounds, cockerel 9 pounds; hen, 8½ pounds; pullet, 7 pounds. Buff Partridge and White Cochins: Cock, 11 pounds; cockerel, 9 pounds; hen, 8½ pounds; and pullet, 7 pounds; Black and White Langshans; Cock, 10 pounds; cockerel, 8 pounds; hen, 7 pounds; and pullet, 6 pounds. The eggs of all of the Asiatic class are a dark brown.

The English Class

The English class is composed of the White, Silver-gray and Colored Dorkings, the Red Caps and the Buff, Black, White, Spangled and Jubilee Orpingtons in both single and rose combs. The White Dorking weighs as follows: Cock, 7½ pounds; cockerel, 6½ pounds; hen, 6 pounds; and pullet, 5 pounds; Silver-gray Dorkings, cock, 8 pounds; cockerel, 7 pounds; hen, 6½ pounds; and pullet, 5½ pounds; Colored Dorkings, cock, 9 pounds; cockerel, 8 pounds; hen, 7 pounds; and pullet, 6 pounds; Red Caps, cock, 7½ pounds; cockerel, 6 pounds; hen, 6 pounds; and pullet, 5 pounds; Orpingtons,

WHITE ORPINGTON, OWNED BY A. W. BESSEY.

cock, 10 pounds; cockerel, 8½ pounds; hen, 8 pounds; and pullet, 7 pounds.

The French Class

The French class is composed of the Houdans, Crevecoeurs, La-Fleche and Faverolles. The Houdans weigh: Cock, 7 pounds; cockerel, 6 pounds; hen, 6 pounds; and pullet, 5 pounds; the Crevecoeurs, cock 8 pounds; cockerel, 7 pounds; hen, 7 pounds; and pullet, 6 pounds. The Crevecoeurs and La Fleche are favorites in France, but are rarely found in this country, as they are not popular in the market here on account of their dark colored shanks.

The Hamburg Class

The Hamburg class is composed of most excellent layers, of white eggs. They are the Silver Spangled, Golden Spangled, Silver Penciled, Golden Penciled, White and Black Hamburgs, and the Silver and Golden Campines. No weights are given for the Hamburgs and Campines.

The Polish Class

The Polish are more of a fancy fowl. They are the White Crested Black, Golden, Silver, White, Bearded Golden, Bearded Silver, Bearded White and Buff Laced. They lay white eggs; no weights are given in the Standard for them.

BEST BIRD AT THE SHOW. SILVER "CAMPINE." COURTESY OF N. E. LUCE.

The Game Class

In the Game class we have the Black Breasted Red, Brown Red, Golden Duckwing, Silver Duckwing, Red Pyle, White, Black and Birchen Games.

PEACOMB BUCKEYE COCKEREL. MRS. FRANK METCALF.

Oriental Class

Are the Cornish, White Dark, and White Laced Red, the Black Sumatras, Black Breasted Malays, and Malay Bantams. The weight for the Cornish is: cock, 9 pounds; cockerel, 8 pounds; hen, 7 pounds; and pullet, 6 pounds.

TYPICAL PAIR BRONZE TURKEYS.

Turkeys

The most popular variety of turkeys is the Bronze; then comes the White Holland, another splendid variety. Among others we have the Black, Buff, Bourbon Red, Slate Narragansett and Wild. The weights for Bronze are: cock, 36 pounds; yearling cock, 33 pounds; cockerel, 25 pounds; hen, 20 pounds; and pullet, 16 pounds; for White Holland, cock, 26 pounds; cockerel, 18 pounds; hen, 16 pounds; pullet, 12 pounds.

Ducks

The Pekin is "The American Duck" with its white plumage and heavily meated body. Their weight is as follows: Adult drake, 8 pounds; young drake, 7 pounds; adult duck, 7 pounds; young duck, 6 pounnds. Another white variety, very popular in England, is the Aylesbury. Weight for adult drake, 9 pounds; young drake, 8 pounds; adult duck, 8 pounds; young duck, 7 pounds. The colored Rouen have similar weights and plumage to the Wild Mallard, the drakes having bright green heads. Other popular varieties are the Indian Runners, both colored and white, called the Leghorn of

GOODACRE'S WHITE INDIAN RUNNERS.

the duck family, being rather small, very active and immense layers of fine white eggs. Then there are the Buff Orpington Ducks, the Blue Swedish, Black Cayuga, Colored and White Muscovy, Call and Black East India, these latter being more ornamental varieties.

Geese

Perhaps the easiest kept and noisiest of all our large variety of domestic fowl are geese, and where conditions are suitable, they prove very profitable. The Toulouse, a large, gray variety, and the White Embden, seem the most popular of the pure-bred varieties, and the weights for either variety are, for adult gander, 20 pounds; young gander, 18 pounds; adult goose, 18 pounds; young Toulouse goose, 15 pounds; and Embden young goose, 16 pounds. Other varieties are the African, Brown and White Chinese, Canadian and Egyptian; these are either used for ornamental purposes or for crossing.

Selection of Breed

Knowing the values and weights of the different standard breeds, the beginner will be enabled to make his choice, and have no trouble in finding the proper selection.

Supposing egg production is the principal object, the beginner will have to decide according to the demand of his nearest market. Boston requires brown eggs, San Francisco white eggs, while Los Angeles seems to be content with either. If you are living near San Francisco, one of the Mediterranean breeds will prove the most valuable to you. The Minorcas, Black Spanish and some of the strains of White Leghorns lay the largest and finest looking eggs. One correspondent who asks for justice for the Minorcas says he has Minorca hens which lay eggs weighing nearly three ounces, and there were Leghorn eggs on exhibition in a late poultry show which weighed five eggs to the pound, but these were from hens "bred to lay." The Brown Leghorns and Hamburgs give many eggs—white eggs also—but smaller, which is an objection in a good market. Should broilers be the object, we should choose the White Wyandottes or White Plymouth Rocks. These latter are exceptionally fine winter layers. For roasters and capons, the Light Brahmas or any of the Plymouth Rocks are the favorites. If two breeds are wanted, we should personally prefer the White Leghorns and White Plymouth Rocks. The White Plymouth Rocks will give the winter eggs and the White Leghorns the spring and summer eggs in great abundance, although they may not lay as many eggs in the winter as the White Rocks. In the early spring the White Rock eggs can be set for early broilers and roasters, while the Leghorns are doing their heaviest laying, and in April and May the Leghorn eggs can be set for the following season's eggs. In this manner there will be a constant succession of eggs for market, and broilers and roasters in season. Always having something to sell means a regular income. Something to market at least once a week. A poultry and egg route and the reputation of having none but the choicest goods to offer is the secret of success.

MERCER'S CHAMPION CORNISH HEN. NOTE THE LARGE BREAST. THE LARGEST BREAST MEAT FOWL.

EGGS FOR BREEDING

Having chosen the breed which suits us best, let us talk on how to get the most out of that breed, for I think we are all agreed that if we keep poultry for profit, we want to make as much as we can out of it. Therefore, having got our fowls, we must treat them right. The natural instinct of a fowl is to make a nest for itself and raise a family of its own in the spring time. It never considers its owner's profit or loss; therefore, to make it answer our purpose, to develop it into a money-maker for us, we must either change its nature or deceive it. We must let it imagine that it is the time of year for nest making and family raising. We must supply it with the conditions of springtime. Our own lives are artificial and the conditions surrounding our domestic hens are also artificial, but we must, if we want success, copy as far as possible Nature's ways with fowls and follow Nature's plans.

EGGS FOR BREEDING. PACKED CORRECTLY FOR SHIPMENT.

In the spring not only do we want egg production, but we want good, strong fertility in our eggs. We want fertile eggs now, for are we not pre-arranging to have plenty of vigorous pullets to lay those high-priced market eggs next fall? Are we not anticipating sturdy cockerels to win prizes at next winter's shows, or to make toothsome fries or delicious roasts?

Fertile eggs are now in order. How shall we get them? First, we must have vigorous and healthy parent birds; we usually have healthy birds in the spring of the year, for the moult is well over and the ailments which prevail in the fall—colds, catarrh and sore throats, all classed as roup—have yielded to treatment, or the victims are no more. The chicken pox, which also is a fall disease, has about disappeared, and the birds are in good condition.

Vigor is Necessary

Vigor is the first requisite for fertile eggs. To have vigor, the hens must have exercise; every grain they eat should be scratched or dug out of the straw or litter in their scratching pen. A hen that is very fat—over fat—will not have fertile eggs and will not have strong, sturdy chickens. It is neither kind nor wise to over-fatten your breeding hens, but they must be fed the proper food for fertility. How can we decide what food to feed for fertility? Let us interrogate Nature again. The wild bird, the Gallus Bankiva, from which sprung all our domestic fowls, lays her eggs and raises her young only in the spring. She only has two broods of about thirteen eggs each, but those eggs are rarely infertile. What does she eat? Principally insects and the tender green grasses or small leaves, not much grain, for the seeds have fallen and have begun to sprout and grow.

During the winter Nature has supplied the birds with grains in plenty, so they have put on fat to withstand the cold; but now there are only a few grains left and the fowls are becoming thinner, yet Nature does not starve them, only gradually changes the ration and gives them worms and larvae, insects of all kinds, for the insect life has also commenced to pulsate and develop; the buds are bursting, too, and the tender green appears and beautiful spring is here, pro-viding all the green food they can eat. How about our captive hens? In our bare back yards, with only the ration we choose to give them? Poor things; they have a natural craving for the tender green, a wild desire for the succulent insect or animal food! See, how they will fight over or scramble for the meat that is thrown to them, or for the head of lettuce! They try to tell us in their own way what they require to produce fertile eggs at this season of the year.

How to Feed

How shall we follow their teachings? Increase the amount of their animal food and give the breeding fowls more green food. How shall we do this? Increase gradually whatever animal food we are now feeding until from 20 to 30 per cent of their daily food is animal food. The best animal food is fresh meat of some kind; the scraps and bones left over at the market; this ground or chopped finely is the best I know of. Rabbits, squirrels, gophers, are all good fresh meat. If fresh meat cannot be obtained, you can get at the poultry supply houses granulated milk, dried blood, blood and bone, beef scrap and other animal food. The best green food is fresh-cut clover lawn clippings, green alfalfa, lettuce, cabbage and other vegetables.

The Male Bird

The male bird is considered as half the pen. The germ or seed of life of the future chicken is from the male. Be sure to have the male vigorous and healthy, and see to it that he gets sufficient food of the right quality. The male bird is often so gallant that he calls up his wives and they greedily eat all the best part of the food, choosing first the meat or animal part, which is the most necessary for fertility, and the husband, the father of future chicks, on which so much depends, is half starved, becomes thin and light. Every male bird when being used to fertilize eggs should be fed extra, either in a pen or corner by himself, or out of your hand at least once a day.

Mating

In mating up the pens I have found the most satisfactory number to mate is about eight or not over ten females of the American breeds to one male. From twelve to fifteen of the Leghorns or Mediterranean birds, and from six to eight of the Asiatic class to one male. Some breeders advocate using two male birds in one pen, alternating them day about, or three male birds for two pens, allowing one bird to rest every second or third day. I never did this, because I was keeping a pedigree of my fowls, and never found any necessity for it.

Caring for Fertile Eggs

Having the fertility assured, the next thing is to take care of the eggs from the time they are laid until incubation begins. Eggs should be kept in a moderately cool, quiet place; not in a draught. I always imitate Nature and turn the eggs, just as a hen would, every day, keeping them in a box either in the cellar or a large, dark, but airy, closet. Some people keep them in fillers with the little end down, but I prefer following Nature's ways and leaving them on their side.

To Choose Eggs for Hatching

To choose the eggs for hatching I use an egg tester or I roll up a newspaper in the shape of a telescope, putting the egg at one end in the sun and my eye at the other end. If the egg shell is speckled or thin at one end, or has thin blotches on it, or is misshapen in any way, or if it feels chalky to the touch, I reject that egg, relegating it to the kitchen, for these eggs will not hatch. I also reject very small eggs, as they are laid by pullets or by over-fat hens, and if they hatch, the chickens will be weaklings. The very large eggs should also be rejected, as they may have double yolks, and these seldom hatch healthy chickens. Above all, never sell for hatching eggs those as described above. The best eggs are the egg-shaped eggs, with good, firm, smooth shells and not narrow waisted.

EGGS FOR MARKET

The hen in her wild state lays about thirty eggs per year. The farmer's average hen lays not over one hundred. On egg farms the average is 150, and some of the fowls of the "bred to lay" strains will average even more.

There are 365 days in the year, and I do not see why a pullet that is fully matured, that comes from an egg-laying strain, a pullet properly fed and cared for, should not lay over 200 eggs per year; in fact, I have had hens that will do even better than that. I will admit that a hen will not lay 200 eggs a year without constant and intelligent care, and the question confronting us is, will the additional number of eggs pay for this care? Also, how shall we give this care and secure these results?

You hear of heredity and pedigree in cows, in horses, in dogs. Heredity is as important with hens as with any other stock. Heredity has as much to do with the success of hens as the right handling. Heredity (or pedigree) and handling must go together. The two-hundred-egg hen must be "bred to lay." She must come from an egg-producing family. No matter how scientifically a hen is fed, or how well housed, you cannot make an extra fine layer out of one whose parents for generations past have been poor layers. It is impossible to take a flock of mongrels and scrubs and get 200 eggs each a year from them, although good handling will greatly increase the yield of even mongrels.

The different breeds require different handling, but no matter what breed you have, there are three essentials to egg production —comfort, exercise and proper food.

Comfort

Under the head of comfort comes first of all cleanliness. A hen that has lice, or fleas, or mites, or ticks on her cannot lay her full amount of eggs. You must help the hen in her efforts to make you money. Give her every encouragement to lay. Cleanliness everywhere. A comfortable, enticing nest, rather dark, where she may stealthily deposit her precious egg. Renew with nice, clean straw once a month. Do everything to coax the hens to lay. If trapnests are used, there should be enough of them so that the hens will not be kept waiting, for by keeping a hen off the nest she will frequently retain her egg until the next day, and will soon learn to be a poor layer. Cleanliness means a clean, sweet-smelling roosting place, where she may sleep undisturbed by lice or mites. Just think for a moment how in the human family a fresh, clean bed in a quiet room will court slumber. I have passed the night in an Arab's tent in Africa that was infested with fleas, and my heart is full of sympathy for a hen that has to live in some of the mite-infested henneries I have seen in the West. Under the head of comfort comes freedom from draughts. A draught in this country will give human beings face ache, neuralgia, earache and a swelled face. It has exactly the same effect on hens. Influenza, swelled head, roup, always or almost always commence from a draught (combined with lice). Comfort means also pure, fresh air without any draught, and pure, fresh water to drink.

Exercise

You know how in the human family exercise is recommended. Physical culture, gymnastics, Ralston exercises, Swedish movements, fencing, etc., and those who may be too feeble to exercise for themselves, pay others to rub, pound and knead or massage them to get the same effect.

Exercise is as necessary for the hen as for the human being and more so, for the hen's exercise of scratching develops the egg producing organs and strengthens them, and hens which exercise lay many more eggs than lazy hens. If you have a vigorous scratcher among your hens, you may be sure she is a good layer.

Exercise a hen must have to develop the egg-making organs. She absolutely must scratch if she is to make a living for herself and you. I consider a scratching pen as necessary for hens in confinement as food. My scratching pens were twelve or fifteen feet long and eight feet wide, but in small yards I have made very satisfactory little pens by nailing four boards six feet long together, forming a square. The boards should be twelve inches wide and the pen filled with wheat straw or alfalfa hay or any good litter. I do not like barley straw on account of the beards, which some times run into the hen's eyes, nostrils, or mouth and cause death. Foxtails, burr clover and wild oats are all dangerous on this account.

I feed all the grain scattered over the straw and my hens scratch and dig happily all day long. The straw or hay is soon broken into short pieces and fresh straw must be added about once a week, and the whole cleaned out and used for mulching trees when the straw becomes dirty. This will depend upon the size of the pen and the number of hens using it.

Proper Food

What it is and how much to give. The scientists tell us that the proper food or the "balanced ration" is composed of one part of protein to four parts of carbo-hydrates. Before discussing this "balanced ration," let us interrogate Nature and find out how a hen balances her own ration.

Let us take a hen as she comes in from foraging in the fields after a long day in summer. Let us kill her and examine her crop. What do we find? Grains of wheat, barley, corn, according to where her rambles have led her; bits of grass, clover and vegetables; some bugs, worms and grasshoppers; here and there a bit of gravel and a lot of matter partially digested that we cannot recognize. The first thing that impresses us is that the hen likes variety, and the second thing that this variety consists of animal food (bugs, worms, insects), grains and green food. This is the "balanced ration," balanced by the hen herself to suit her needs in the summer time when eggs are plentiful. If we want eggs in the winter, we must, as far as possible, give the same conditions, the same variety of foods, with plenty of pure, fresh water, never forgetting that about seventy per cent of the egg is water.

But to return to the "balanced ration." We know that a hen requires a certain amount of food to keep her alive and thriving; above that the surplus goes either to making the egg inside her or to making fat.

The hen is an egg-making machine, but if you put into that machine none of the elements of the egg, you cannot expect the machine to turn out eggs.

Therefore, the scientists analyzed the egg, and not only that, but also analyzed the body of the hen with the feathers, and discovered as follows: The very large number of different substances found in the hen may be grouped under four heads: 1, water; 2, ash or mineral matter; 3, protein (or nitrogenous matter); 4, fat. The proportion of each of these groups alters with the condition of the hen. Water is the largest ingredient and amounts to from forty to sixty per cent of the weight of the bird. Ash or mineral matter forms from three to six per cent when the hen is not laying, and from six to ten per cent when laying. The group called protein constitutes from fifteen to thirty per cent of the weight. Fat seldom falls below six or rises above thirty per cent.

The feathers are composed of protein and ash, the ash being largely silicate of potash and lime.

The accompanying analysis of the hen, pullet and egg has been kindly sent to me by Professor Jaffa; that of the egg was made by him at the University Laboratory of California.

Analysis of Hen and Egg

	Typical Leghorn Hen	Pullet in full laying, Leghorn	Capon, Plymouth Rock	Eggs as Purchased	Eggs, edible Portion
Water	56.8	57.4	41.6	65.6	73.7
Protein	21.6	21.2	19.4	11.8	13.3
Ash	3.8	3.4	3.7	.7	.8
Fat	17.8	18.0	35.3	10.8	12.2
Shell	...	...	....	11.1	...
Total	100.0	100.0	100.0	100.0	100.0

Composition of Hen and Egg

Calculated on a Water-free Basis

Protein	50.0	49.8	34.3	50.5
Fat	41.2	42.2	31.4	46.4
Ash	8.8	8.0	2.1	3.1
Shell	...	...	32.2	...
Total	100.0	100.0	100.0	100.0

It is interesting to compare the analysis of the hen and egg with some of our grains and poultry food. In all our grains are found more or less the elements of the egg, but they are not in the right or proper proportion for making the egg. There is usually too much of the fattening element in the grains and not enough protein or nitrogenous element, which forms the meat, muscle, bone and feather. This is the most valuable and most expensive part of the ration.

In order to keep up the strength of the hen and have her produce the largest amount of eggs, it has been found that for every pound of protein in the food, she must have four pounds of carbo hydrates. This will vary slightly according to the heat of the weather and the needs of the hen.

I would urge you to send a postal to the University of California at Berkeley, asking for the Farmer's Bulletin No. 164 on Poultry Feeding. This bulletin, by Professor Jaffa, is one of the most valuable bulletins ever published. It contains the analysis of the different grains, vegetables and meats and of most of the proprietary foods, besides formulas for the best rations.

In Bulletin 140 of the Department of Agriculture there are some rules for caring for eggs for market which are good:

1. Use hens that produce not only a goodly number of eggs, but those of standard size. Such breeds are Plymouth Rocks, Wyandottes, Rhode Island Reds, Leghorns, Orpingtons and Minorcas. 2. Good housing, regular feeding, and, above all, clean, dry nests. 3. Daily gathering of eggs, and when the temperature is above 80 degrees, gathering twice a day. 4. Confining all broody hens as soon as they show symptoms of broodiness. 5. Rejection of all doubtful eggs found in a nest that was not visited the previous day. 6. Placing all summer eggs when gathered in the coolest place available. 7. Prevention at all times of moisture coming in contact with the eggs. 8. Disposing of young cockerels before they begin to annoy the hens. 9. The using of cracked and dirty eggs at home. 10. Marketing all eggs at least once a week or oftener. 11. Keeping all eggs cool while on the way to town or in the country stores. 12. Keeping all eggs away from bad odors and out of musty cellars. 13. The use of strong, clean egg cases and good fillers.

SPROUTING OATS

By W. S. Willis

The following method of sprouting oats has been kindly sent to the author by Mr. W. S. Willis, of the celebrated Arlington Egg Ranch. Mr. Willis has found the sprouted oats a splendid addition to the hen's ration, lending variety to the daily bill of fare and increasing the egg output.

Three quarts of oats will make a fine morning meal for 100 hens if properly sprouted.

Place the grain in a pail and let it soak for twenty-four hours; then transfer it to a box one foot square and six inches deep, with a few small drainage holes in the bottom.

Sprinkle with water daily and allow the grain to remain in the box until the sprouts are from two to three inches in length, at which time it will be ready to feed.

As it takes from eight to ten days to secure the proper growth, a number of boxes or compartments should be provided for the grain, keeping each day's allowance separate, and a new lot should be started daily.

For larger flocks of course it is necessary to increase the size of the boxes—a day's feed for 600 hens, for instance, requiring a sprouting space of two by three feet.

In all cases care should be taken not to have the grain over two inches deep when placed in boxes, in order to guard against heating and mildew.

The boxes should be placed in a level position and kept covered with a board or burlap, in order to keep the grain in a moist condition.

In cold weather the sprouting operations should be conducted in comfortably warm quarters, and warm water may sometimes be used to advantage in sprinkling the grain.

Redwood is better than pine to use in making the sprouting boxes, being less liable to swell and crack when water soaked.

Should it be impossible to get oats that will grow well, barley or wheat may be substituted, but it may be found necessary to stir the barley until it begins to sprout, to prevent fermentation.

THE FEEDING PROBLEM

The three essentials of egg production, the three essentials of profit in poultry keeping, the three essentials for vigor and health in fowls are—comfort, exercise and proper food.

Let us consider (1) the proper food, (2) the methods of feeding it, and (3) recipes for a few tried balanced rations.

Practical knowledge and skill in feeding can be acquired without the study of science. Feeding fowls for good results is a comparatively simple matter.

Requirements in Feeding

The food which a fowl consumes has three chief functions to perform: (1) to sustain life, promote life, repair waste and produce eggs; (2) to keep the body warm; (3) to furnish strength or energy which is expended in every movement. The fowl is also able to store food, not needed at the time it is eaten, for future use; this store is chiefly in the form of fat, which serves as a reserve supply of fuel.

Food Elements

To supply the three functions in the life of a fowl there are three principal food elements: Proteins, carbo-hydrates and fat; all of these are contained in the different grains and foods used for poultry.

(1) Proteids (or protein), albuminous or nitrogenous matter. Protein is the nourishing matter, the principal tissue former, supplying material for bone, muscle, blood, feathers, eggs. Its latent energy can also be converted into heat and energy; but it is more costly for such purposes than the non-nitrogenous foods.

(2) Carbo-hydrates, carbonaceous matter, starches and sugar. Carbo-hydrates form the bulk in nearly all foods and are the principal sources of heat and energy.

(3) Fats are found in almost all foods. They furnish heat and energy in addition to the supply from the carbo-hydrates. Fat also enters largely into the composition of the yolk of the egg.

All three food elements are necessary. The proper combination of these three is called the "balanced ration." It is, in other words, a "complete" ration, containing in proper proportions the necessary food elements to promote (1) growth, including egg production, (2) warmth, and (3) energy or strength. The needs of a fowl's system are not always the same; it does not always need the different elements to be in the same proportions; the ration properly balanced or suitable for a growing chick would be unbalanced (unsuitable) for the mature hen. The food to be a balanced ration must be adapted to the present needs of the fowl.

Many people find it easier to keep food values in their minds when they have seen a picture than after studying over figures in a table. A glance at a couple of foods to be compared, with the proportion of ingredients blocked out plainly, as they are here, makes an indelible impression on the mind. One can see in a moment where one pays for water in the foodstuffs and where one does not. When it is desirable to know the exact percentage of protein or carbo-hydrate that a food contains, it is necessary to refer to the table for ease of calculation.

Material	Water	Ash	Protein	Carbo-hydrates. Starch, Fiber	Sugar, etc.	Fat
Milk	87.20	.70	3.60		4.90	3.70
Skim Milk	90.60	.70	3.30		5.30	.10
Dried Milk	12.10	15.10	58.80		12.40	1.60
Cottage Cheese	72.00	1.80	20.90		4.30	1.00
Fresh Meat	71.00	1.00	22.00		....	7.00
Beef Scraps	5.00	17.00	59.00	3.40	3.85	17.00
Cocoanut Oil Cake Meal	14.08	4.36	19.51	9.53	42.12	10.40
Linseed Oil Cake Meal	10.93	4.50	30.70	8.89	37.95	7.03
Cotton Seed Meal	9.85	4.86	47.25	3.19	22.64	12.21
Soy Bean Meal	9.50	5.60	44.40	4.35	28.44	7.70
Gluten Feed	7.80	1.10	24.00	5.30	51.20	10.60
Beans, dried	12.60	3.50	22.50	4.40	55.20	1.80
Peas, dried	9.50	2.90	24.60	4.50	57.50	1.00
Barley, rolled	10.05	2.92	12.00	2.30	69.63	3.12
Barley, sprouted	55.50	1.18	7.00	4.26	31.14	.75
Oats	11.00	3.00	11.80	9.50	59.70	5.00
Oats, rolled	7.70	2.00	16.00	1.30	65.00	7.00
Corn, Indian	10.60	1.50	10.30	2.20	70.40	5.00
Rice	12.30	.30	8.40	78.60		.40
Rice Bran	10.55	6.64	14.96	4.85	50.20	12.80
Rye	11.60	1.90	10.60	1.70	72.50	1.70
Wheat, plump	11.50	1.76	11.85	2.45	70.40	2.03
Wheat, shrunken	8.30	2.34	17.10	3.48	66.78	3.00
Wheat, bran	11.67	5.18	14.05	8.16	57.34	3.60
Wheat, middlings	11.73	2.85	15.22	4.88	60.85	4.47
Wheat, shorts	9.85	4.24	15.20	5.05	64.48	3.32
Mixed Feed	10.57	3.57	12.00	9.66	59.98	4.21
Broken Crackers	5.90	1.90	10.00	.80	70.30	9.00
Cabbage	90.50	1.40	2.40	1.50	3.90	.40
Alfalfa, green	80.00	1.72	4.94	4.70	7.90	.74
Alfalfa, meal or hay	10.95	6.43	17.60	22.63	39.31	3.08
Pumpkins	90.90	.50	1.30	1.70	5.20	.40

Methods of Feeding

The question of how to feed and what to feed for the best results in egg production is the most difficult problem in poultry keeping, and has for some time been engaging the attention of the various Government Experiment Stations in this and other countries. The two successful systems in use at the present time are the Mash system and the Dry Feed system.

The mash system is one in which a mash is fed once or twice a day. The foundation of the mash is bran, middlings, and corn meal or chops. It is mixed wet, raw, scalded or cooked. The dry feed system is when a dry mash is fed, consisting of the same ingredients as the wet mash, but dry. Dry feeding is used by many regularly, and is becoming more popular every year.

In mash feeding the errors to be avoided are: Too concentrated a mash with too much meat or fat; too light or bulky, that is,

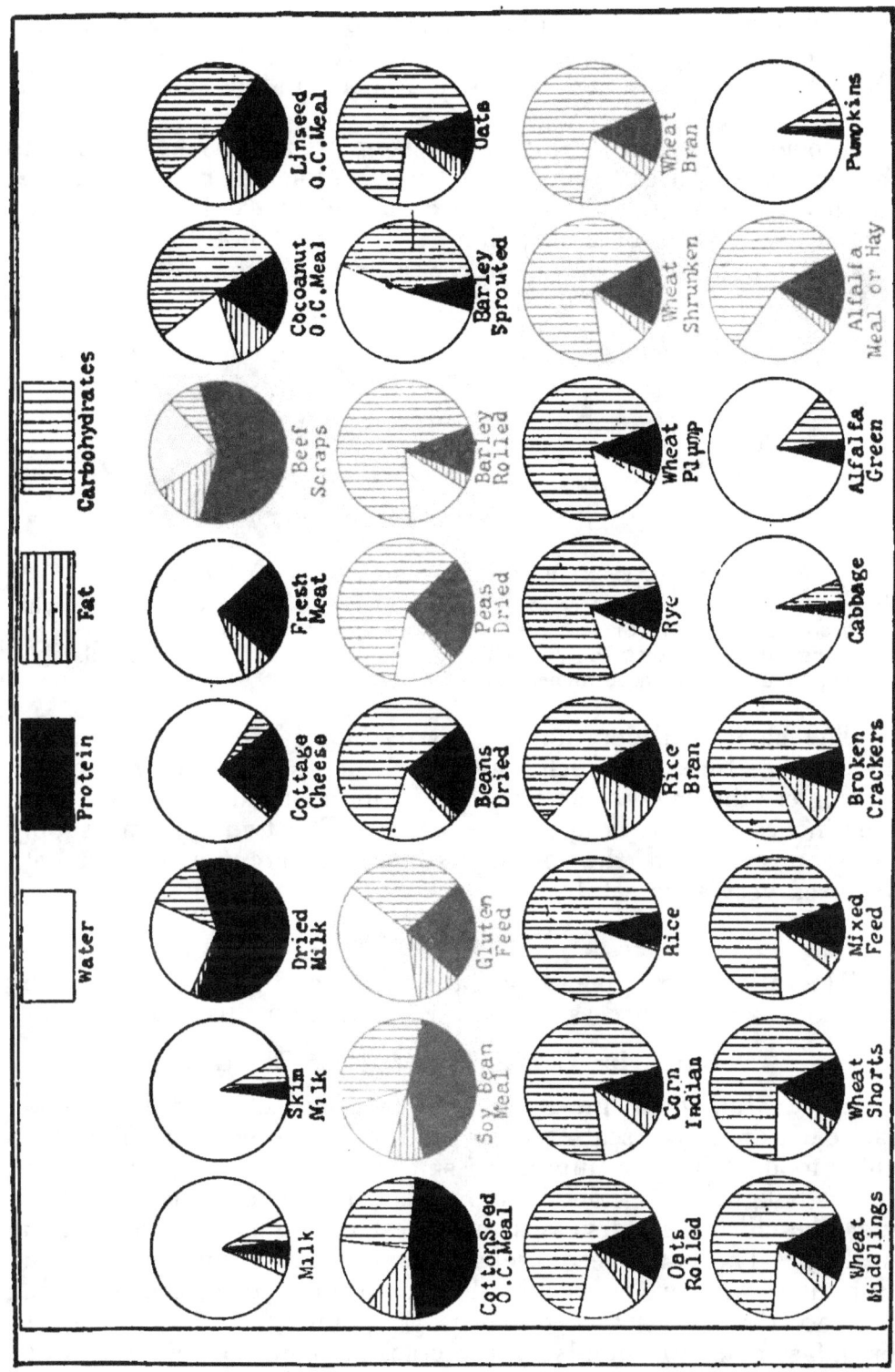

By courtesy of the University of California Experiment Station.
See table on opposite page.

composed principally of bran or hay; too wet or sloppy or sour or mouldy. Experience has shown that feeding wet mashes more than once a day has bad effects, producing indigestion in various forms.

The advantages of the dry-feed system are: A saving of labor to the feeder, is lighter to handle and much easier to mix. It can be fed in the morning. The fowls are obliged to eat it slowly; they cannot swallow it in a few minutes. It will not freeze in cold weather nor become sour in hot weather, and the fowls will not over-eat with the dry feed.

AN EXCELLENT FEED HOPPER FOR YOUNG AND OLD.

These hoppers are made 8 feet long. The trough is 8 inches wide and 4 inches deep with a strip (or lath) half an inch wide nailed along the top of trough inside to keep the chickens from pulling out the feed. The slats are about two to three inches apart.

The chief consideration in dry-feeding is that fowls require about three times as much water to drink as with the wet mash; also unless the dry food is placed in hoppers or fed in boxes at least four inches deep, it is apt to be wasted. The two systems supply the requirements of the fowls in slightly different ways and both are used very successfully.

Sample Rations

The rations here given have been tested and proved excellent by some of the most successful poultry breeders in this country.

Ration for Chicks Intended for Breeders

First meal, when chicks are 36 hours old: Rolled or flake breakfast oats, dry; give scattered on sand every three hours, then feed chick food. This is a number of small or broken dry grains which can be bought at the poultry supply houses. The use of hard grain diet like chick feed, develops the digestive organs and keeps them healthy. The chick feed prepared by reliable firms is excellent. For those who prefer to mix their own chick feed, the following is a good recipe: Cracked wheat, 30 pounds; steel-cut or rolled breakfast oats, 30 pounds; finely cracked corn, 15 pounds; millet, rice, pearl barley, rape seed, finely ground beef scraps or granulated milk, dried granulated bone, chick grit, 10 pounds; granulated charcoal, 5 pounds. In the chick feeds wheat, oats and corn are the staples, the most necessary part of the ration. Feed at 6 a.m. chick feed scattered in chaff; 9 a.m. rolled or steel-cut oats; 11 a.m. green lettuce; 1 p.m. chick feed; 3 p.m. green feed; lettuce, clover or potatoes chopped fine; 4:30 p.m. hard boiled eggs (4 for 100 chicks), chopped shell and all, with the same amount of onions and twice the amount of bread crumbs or rolled oats or Johnny-cake. One fountain of skim milk and one of clean water always before them and renewed three times a day. Very coarse sand and granulated charcoal should be always before them.

Toward the end of the second week mix a little whole wheat, hulled oats and kaffir corn with the chick food, gradually increasing it, until at the end of the sixth week they will be eating this entirely.

Rations for Broilers

For the first two weeks use the same feed as given for the breeders. Third week, 6 a.m. chick feed; 9 a.m. mash, 1 part each of bran, cornmeal and rolled oats, and a little salt; mix with skim milk, making a crumbly dry feed in a small dish or trough, taking away all there is left in fifteen minutes; 11 a.m. lettuce or clover; 1 p.m. rolled oats; 3 p.m. chopped raw potatoes; 4:30 p.m. mash same as in the morning. Fourth week, 6 a.m. chick feed; 9 a.m. mash, adding 5 per cent beef scraps or cracklings; 1 p.m. chopped potatoes; 4:30 p.m. mash, same as in the morning. Keep grit and charcoal always before them, with skim milk and pure water. Finish off at six to eight weeks by gradually adding from five to ten per cent of cotton-seed meal and a little molasses with the mash.

Rations for Laying Hens

In order to keep up the strength of the hen and have her produce the largest amount of eggs, it has been found that for every pound of protein in the food she must have four pounds of carbo-hydrates. Many instances may be cited in which the rations fed to laying hens differed greatly, but have been productive of excellent results, provided they contain a sufficient quantity of digestible protein. The following rations have proven successful:

I will give a formula that I have used for many years after experimenting with others, and will give some that are being used at the present time by prominent and successful breeders near here. There are many other breeders, but I happen to have these by me and have not those of the others. The Basley formula is as follows: By measure, 2 parts heavy bran, 1 part alfalfa meal, 1 part corn meal, 1 part oatmeal (called Breakfast Flaked Oats), 1 part beef scraps or meat meal or granulated milk, a little pepper and salt; keep this in a hopper or feed box. At noon green feed. In the evening grain, wheat, kaffir corn or cracked corn, barley, hulled oats, equal parts, mixed and scattered in straw in the scratching pen. Fresh water constantly before them; if they run out of water, the egg yield will stop. I keep before the fowls at all times sharp grit, crushed oyster shells, charcoal and granulated dried bone. At moulting time I add to the grain sunflower seed, and to the dry mash linseed meal. The reason I feed oatmeal is that I always feed for vigor. I want the parent birds to be vigorous and the eggs to have such an amount of protein in them that the chicks will not fail in being vigorous. There is no food equal to oats for giving vigor. The reason I feed alfalfa is that although it shows on analysis almost the same protein content as bran, it gives the yolk of the eggs a rich orange hue which bran fails to impart. All fowls need plenty of green food and clean water. The green food is the cheapest food you can give and keeps the digestive organs in good condition. Green food must be given daily.

Rations of Successful Breeders

Wilcox Standard Mash—50 lbs. heavy wheat bran, 20 lbs. corn meal, 14 lbs. ground barley, 5 lbs. oil cake or cotton-seed meal, 10 lbs. beef scrap, 1 lb. fine charcoal.

Johnson Formula—80 lbs. wheat bran, 15 lbs. alfalfa meal, 15 lbs. cracked raw bone, 1 pint of home-made condiment.

Bickford Dry Mash—One part corn meal, 1 part middlings, 2 parts heavy wheat bran, 1-10 part meat or blood meal, 1-10 cotton-seed meal, a good handful of salt to one hundred pounds.

Goodacre Standard Mash—Ten lbs. wheat bran, 2 lbs. corn meal, 2 lbs. fine meat meal, 1 lb. linseed meal.

Walton's Dry Mash—12 parts wheat bran, 4 parts corn meal, 2 parts beef scrap, 2 parts alfalfa meal, 2 parts granulated milk, ½ part charcoal.

Cowles Dry Mash—One part each of corn, wheat and barley ground up together. To 80 lbs. of the above add 5 lbs. of blood meal, 5 lbs. of bone meal, 10 lbs. of meat meal and a little charcoal.

For One Dozen Hens

Rations for one dozen breeding hens, American class, in confinement, for three days' rotation.

Monday morning—One pint and a half grain, wheat, cracked corn and hulled oats, equal parts mixed and scattered in straw or litter in scratching pen. Noon: Cut clover or lawn clippings. Evening: Mash, 1 qt. heavy bran; 1 qt. ground oats; 1 pt. corn meal; 1-3 of the whole cut clover or alfalfa meal; 1 tablespoonful each of salt and pulverized charcoal; ½ pt. beef scraps.

Tuesday morning—1½ pts. mixed grain, wheat and rolled barley. Noon: green feed, pumpkins or clover; 1 pt. green cut bone. Evening: Mash, 1 pt. cooked vegetables and table scraps, 1 qt. bran, 1 pt. cornmeal, a little salt and pepper.

Wednesday morning—1½ pt. mixed grain; wheat, hulled oats, kaffir corn. Noon: Cabbage or beets. Evening: Mash, 1 pt. peas or beans soaked over night, boiled with a little soda until soft; ½ pt. dried blood, or beef scraps, 1-3 cut clover. If you cannot get beans cheaply, use potatoes or other vegetables.

Follow the same system the remaining three days.

Sunday, instead of the mash, scald three pints of rolled barley in the morning, cover and leave to steam. Feed in the evening instead of the mash; this makes a pleasant change and saves work for the Sabbath.

The reason for feeding the mash at night is to keep the hens busy scratching all day and so send them to roost with their crops full. There is danger of the American and Asiatic fowls becoming too tat and lazy without exercise if given the mash in the morning.

Bulletin 164 of the California College of Agriculture gives the following formulas as samples of the many different combinations that can be made from the various feed stuffs on the market. They are calculated for 100 hens a day, and if fed with nine to twelve pounds of grain, according to weight of hens, and some green stuff, they will form a well balanced ration. Mash may be fed wet or dry.

I.

	Quarts	Pounds
Bran	6.0	3.0
Shorts	2.5	1.8
Corn meal	1.5	2.3
Cocoa O. C. meal	1.0	0.9
Beef scrap	1.0	1.5
Coarse bone meal	0.5	1.0

II.

	Quarts	Pounds
Bran	6.0	3.0
Middlings	0.5	0.5
Linseed O. C. meal	0.5	0.5
Gluten feed	0.8	1.0
Ground oats	1.0	0.75
Corn meal	1.5	2.25
Beef scrap	1.0	1.50
Coarse bone meal	0.5	1.00

III.

	Quarts	Pounds
Bran	5.0	2.50
Shorts	3.0	2.00
Corn meal	1.5	2.25
Soy bean meal	0.75	1.00
Beef scrap	1.00	1.50
Coarse bone meal	0.50	1.00

IV.

	Quarts	Pounds
Bran	6.0	3.0
Corn meal	1.0	1.5
Barley meal	2.0	2.2
Alfalfa meal	1.0	0.5
Soy bean meal	1.0	1.3
Beef scrap	1.0	1.5
Coarse bone meal	0.5	1.0

V.

	Quarts	Pounds
Bran	4.0	2.0
Alfalfa meal	1.0	0.5
Corn meal	1.0	1.5
Shorts	2.0	1.5
Barley meal	1.0	1.1
Ground beans	1.0	1.1
Beef scrap	1.0	1.5
Coarse bone meal	0.5	1.0

VI.

	Quarts	Pounds
Bran	5.0	2.5
Alfalfa meal	1.0	0.5
Corn meal	1.5	2.3
Linseed O. C. meal	1.0	0.9
Shorts	2.0	1.5
Beef scrap	1.0	1.5
Coarse bone meal	0.5	1.0

Salt should be added to every mash, about an ounce being sufficient. Pepper may be added occasionally. Fresh lean meat may be substituted for beef scrap in any of the formulas, three quarts of the fresh being equal to one of the dried. Cottage cheese may be substituted in the same proportion, except that it is advisable not to replace all of the meat, one-half quart beef scrap and one and a half quarts cottage cheese being a much better proportion. The equivalent in pounds is given for convenience in ordering. The quarts represent the amount for 100 hens and may be multiplied or divided ad libitum.

Fattening Fowls

Fowls to be fattened should be confined in small yards or in coops or crates, especially adapted for feeding. The object in keeping them in confinement is to prevent the forming of muscle and sinew, which would occur if allowed to run at liberty.

The crate used for fattening fowls can be four or six feet long. Mine were composed of lath six feet long; the frame of the crate is 6 feet long, 18 inches wide and 18 inches high, divided into six little stalls or compartments. The frame is covered with lath, placed lengthwise on the bottom, back and top the width of one

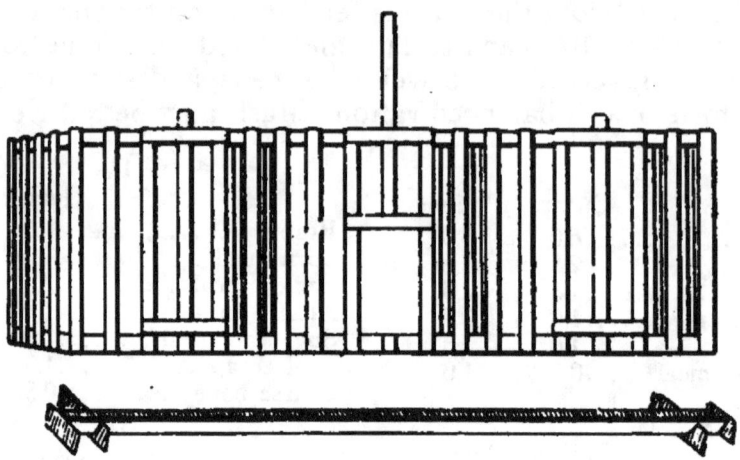

THREE-COMPARTMENT FATTENING COOP AND TROUGH.

lath apart. The first lath on the bottom should be two inches from the back to allow the droppings to fall through, otherwise they would lodge on the lath at the back. The lath are placed up and down in the front, the spaces between them being two inches wide to enable the chickens to feed from the trough. A "V" shaped trough is made to fit into two notches in cleats in front of each crate. The crate stands 15 inches from the ground; the droppings are received on sand or other absorbent material and removed daily. The coop is large enough to hold 12 or 18 young chicks (2 or 3 in a stall) or six full grown fowls. Fowls are fed three times a day all they will eat in 15 minutes.

See cut of fattening crate.

Formulas for fattening:

(1)   Equal parts of bran, cornmeal and oat meal (rolled breakfast oats) mixed with skim milk, fed three times a day.

(2)   Buckwheat flour, pulverized oats, cornmeal in equal parts, mixed thin with buttermilk.

(3)   Equal parts barley meal and oat meal and a half part of corn meal, mixed with buttermilk or skim milk.

(4)   A favorite French combination is two parts barley meal, one part cornmeal, one part buckwheat flour.

A little salt and coarse sand should be added to their food. Three weeks is the length of time to continue the feeding. Chickens do not seem to be able to stand the confinement for a greater length of time. The last week of the fattening process, five per cent of cotton seed meal and a little tallow may be added to any of the above formulas.

Feeding Beans

Our readers know our "Rule of three"—or the three essentials of egg production—Comfort, Exercise and Proper Food—and how very necessary each of this trio is for filling the egg basket.

The successful poultry breeders, those that are really making money in the poultry or egg business, all and each follow our Rule of three. Some put more emphasis on one of the three conditions, and some on the other, but I find the man that uses all three essentials about evenly balanced is the successful man.

Just at present there are several of our readers who are seeking for advice on the problem of the proper food and have appealed to me for information about the use of beans and some other foods which are available or cheap in their locality. I would like to help them discuss this subject together with the different breeds they are feeding.

We all know that food is first necessary to sustain life, to enable the young fowls to grow and make their feathers, while it also enables the mature fowls to make and produce eggs. We have learnt that the body of the hen and the egg also is composed of water, mineral matter, nitrogenous matter and fat, and that to sustain life and growth and to produce eggs, the hen must be supplied with these elements. It is exceedingly interesting to learn the right proportion of these different elements that have to be supplied to the hen, all of which may be found in the analysis of the different foods given in the valuable bulletin "Poultry Feeding and Proprietary Foods," by Professor Jaffa of the University of California.

Professor Rice of Cornell, in one of his lectures, says, "Feeding poultry is a science and an art." The science is in the knowing why, and the art is in the knowing how to do it. Our Professor Jaffa divides the food (this is the science part) into three classes: The protein, carbo-hydrates and fat. He explains that the word protein comes from a Greek word which means the chief thing— or the first thing—and the protein is the most important part of the food, for by it is made or produced the bone, muscle, blood, nerves, tendons, etc. The protein or nitrogenous matter of the hen's body and of the egg is formed by the nitrogenous matter (the protein) that is fed to the hen or that she finds in hunting on the range for her food, so anyone can see how important this element is in the food.

The carbonaceous part of the food, which includes the fat and carbo-hydrates (sugar and starch), is mainly used as a fuel supply to the body and is the substance which is consumed in the production of heat and energy. We know or have learnt that an active fowl, such as a Leghorn that is always on the move, scratching, running, flying, uses up more of the fat-producing food than a quieter, tamer, heavier fowl, such as the Plymouth Rock or Wyandotte or one of the Asiatics.

The scientists have analyzed the food as well as the hen and have decided that a hen requires as a balanced ration for egg production one pound of protein to four pounds of carbo-hydrates, and we believe this and act on it by giving the hens animal food, green food and grain. We also want to get the food as cheaply as possible to save our pocketbooks, and yet give the hens food that will bring the best results, this is usually eggs when eggs are dearest.

The protein is the most expensive part of the food, consequently when we find a food that is inexpensive but contains a large amount of protein, we are glad to buy it, and then we must find out how to mix it or with what other food in order to get the right balance of one part of protein to 4 or 4.5 of carbo-hydrates. A ration means the food for a whole day.

I am always glad to talk over the different foods and to help beginners decide what is the best and cheapest food for them to use in their locality. Several have lately asked about BEANS, how to feed them to the best advantage. Some years ago I had an opportunity of buying a large quantity of navy beans that had been held as seed beans but several sacks of them had become weevily. I studied Professor Jaffa's bulletin and decided that it would be a good plan to buy them, thinking that as they were small, the hens would eat them, but my hens did not take to them at first, so I sent the beans to the mill and had them coarsely ground, and I then soaked them over night with a little bicarbonate of soda in the water, and the next morning when the fire was lighted for breakfast, I put on the beans and let them cook at the back of the stove, taking them off at noon and mixing in bran and cornmeal, also a little alfalfa meal, and seasoning with salt and pepper as for the table. The hens like this mash made of bean soup, and never hens laid better than these. It was certainly a famous egg food.

Recently I received letters from several of our readers asking about feeding beans, and I replied, giving Professor Jaffa's analysis, but I afterwards received a letter asking me for the analysis and the value of "broad Windsor beans," and as there was no analysis of them in the bulletin, I sent some of them to the Agricultural College to have them analyzed. Professor Jaffa not only analyzed them, but also analyzed some "horse beans," as I said that Windsor beans were sometimes called horse beans and were largely fed to horses in some places. The horse beans that he bought were larger than the Windsor beans that I sent him and he found both of them so exceedingly rich in protein that, to be certain there was no mistake, he had the analysis duplicated, done over twice.

Analysis of Horse Beans

	Per cents
Water	14.05
Ash	2.10
Protein	25.10
Fat	1.60
Fiber	6.63
Starch, etc.	50.52
Total	100.00

Analysis of Windsor Beans

	Per cents
Water	10.98
Ash	3.02
Protein	18.80
Fat	1.58
Fiber	6.65
Starch, etc.	58.97
Total	100.00

Analysis of Navy Beans

	Per cents
Water	12.60
Ash	3.50
Protein	22.50
Fat	1.80
Fiber	4.40
Starch, etc.	55.20
Total	100.00

It will be seen by these analyses how rich in protein are the beans, and, therefore, what a valuable food for fowls. Realizing the value of this, in order to help other of our readers, I wrote to A. A. W. for further information about the beans he had sent me, and received the following reply:

"The beans are commonly known in England (where they are very popular) as 'broad Windsor beans,' but to the best of my remembrance these are a smaller species. I raised these here on rich soil apparently high in nitrogen, judging by the rank top growth of various crops planted therein; the vines averaged a height of over seven feet, which is more than double that claimed for them by the seedsmen, who do not usually underestimate the vigor and prolificacy of their well-advertised goods. I have a copy of your poultry book and believe I have derived much profit from it, as I am raising broilers and feeding them entirely according to your directions*; some of them weigh close on to two pounds each, and none of them are over six weeks and four days old, raised in brooder coops without hens or artificial heat, but with the best possible care and attention to details, and with less loss than I expected, as this is my first experience of this way of raising them. May I trouble you to inform me of the best method of feeding the beans to chicks of various ages, as I have others at different stages. I have fed them occasionally to month-old chicks in small quantities by soaking until the skins will slip, then chopping up fine with bran to make a crumbly mash. I would much like to know if this is a good combination or otherwise, and how best and when to feed, and the proportion of beans, and whether chopped up dry, soaked or cooked.

"My idea in discarding the skins is that being very tough and leathery, they might possibly be indigestible."

In reply to this, the skins are very tough, that is, the skins of both horse beans and Windsor beans, and it was a wise precau-

---

* See Rations for Broilers, page 39.

tion to take them off for the little chicks, but that would scarcely be possible or profitable if you are feeding much to mature hens, as it would take too much time and labor.

In feeding either old or young you can make one-fifth of the food of the beans if you have plenty of them, but I would advise not more than that. Your way of mixing the chopped-up beans with bran and milk is good, but I would suggest adding a little cornmeal, about one-fifth of the amount of the mash. This would be a better balanced mash. As you have had such good results from following my instructions and formula for feeding broilers, I think you had better continue it and not make any change, or if for any good reason you are obliged to make a change in the food make the change very gradually; that is, add only a few spoonsful of the new food each day until at the end of about two weeks you have got them to willingly accept the new food. A sudden change of almost any kind will stop the egg output partially or sometimes totally. You have to remember there is a difference between variety, which is excellent for fowls, and change, which almost invariably results disastrously.

The best way to feed the beans (Windsor or horse beans) would be to have them ground and feed them in the dry mash for all the chickens, large or small; for the very little fellows nothing could be better than the way you are now doing.

When I received this letter I wrote to a successful poultryman and egg farmer, who has been feeding beans for some time very successfully, and I copy his letter for the benefit especially of those residing in bean-growing districts, where beans can be often bought very cheaply. The writer can be thoroughly relied upon as to accuracy.

"Your cordial letter reached us today, and I take pleasure in answering your questions concerning our use of beans for hens.

"The variety we used and are still using is what is called here the black-eyed bean. I think it is called 'cow peas' in some parts of the country. The flavor of this bean is more like that of the pea than of the bean. For a long time we fed them whole, with corn, wheat and whole barley, equal parts of each. The hens ate them as readily as they did the other grains, except wheat.

"We fed it also in the mash, with ground barley, cornmeal and beans, about equal parts of each. We found that our hens increased their egg production about twenty per cent.

These beans are rich in protein, about 22%, and are about 85% digestible, so you will see that fed with wheat, corn and barley they are a valuable addition to the dietary of hens. If we could get these beans, we should continue their use, but we are unable to get any more of them. If you know where they can be had for a reasonable price, we should be pleased to have you inform us. I have no doubt that hens could be induced to eat lima beans, at least in the mash, as you know lima beans are rich in protein, but possibly may not be as digestible as the black-eye. I hope this information may be of use to you."

In this article we give the scientific side, the analysis of three kinds of beans, and also the practical use of them by three different poultry breeders. This will answer several other inquiries on the subject, and we hope prove useful to many of our readers.

Feeding Alfalfa

Alfalfa is one of the most valuable of green foods for both cattle and poultry. I have found by my own experience that what is good for the milch cow is good for the laying hen. You know the Holstein does best on a large amount of succulent food, while the smaller Jersey is adapted to a more concentrated diet; so with the different breeds of hens, the Leghorn needs a wider ration than the Plymouth Rock, the ration that would bring the largest amount of eggs from the Leghorn would so fatten the Rock that it might prevent her laying.

Where alfalfa is abundant the following ration will give good results. Alfalfa hay or meal when good is rich in protein and resembles wheat bran in analysis. In the last analysis of alfalfa hay or meal Professor Jaffa gives protein 12.3, carbo-hydrates 37.1, fat 1.6; while wheat bran has protein 12.62, carbo-hydrates 38.88, fat 2.5. By this you see that whilst the protein content of alfalfa meal is almost the same as wheat bran, the fat, sugar and starch is decidedly lower, also the crude fiber, which is indigestible, in alfalfa meal is 22.63 per cent, whilst in bran it is only 8.60 per cent, to counterbalance this we can use more corn meal or some linseed meal. While alfalfa is rich in protein, it has been found by experiment that the fowls need animal protein, as substitute for bugs and worms, so to the ration must be added animal food in some shape, usually beef scrap,, meat meal or milk. Cottage cheese is one of the best animal proteins.

The following ration will prove satisfactory: Alfalfa meal, 50 lbs.; corn meal, 20 lbs.; barley meal, 10 lbs.; beef scrap, 10 lbs.; coarse bone meal, 10 lbs.

I gave my Indian runner ducks a similar ration: 30 lbs. alfalfa meal, 30 lbs. heavy bran, 30 lbs. corn meal, 10 lbs. beef scrap; giving at the same time all the crushed clam shells they would eat. On this ration ducks averaged 233 eggs each per year, and kept in perfect health.

I wrote to Mr. Hammonds, who is manager of the largest broiler plant in the country and an expert poultryman, a graduate of Cornell, and asked about a pullet that he was bringing up on nothing but alfalfa and milk. His answer is as follows:

"The pullet you refer to laid 253 eggs from August 7, 1910, to August 7, 1911, and she gained one and a quarter pounds in weight. All of her eggs, except the first six, weighed two ounces or more and were of good shape and shell. She was fed from the time she was two months old until she was five months old and showed signs of laying on alfalfa meal moistened with milk, and rolled barley as ·grain. From the time she was five months old till the test was finished she was fed a dry mash composed of the following: Alfalfa meal, 3 lbs.; barley meal, 2 lbs.; bran, 1 lb.; corn meal, 1 lb., and fine beef scrap, 1 lb. As grain, she received one handful of rolled barley at 4 p.m. each day. Aside from this she had dry alfalfa hay as a litter. After she began laying she was often tempted with green alfalfa but did not seem to care for it, therefore she was not fed any green stuff".

"Champion JUMBO" 1st. Prize C'k'r'l. CHICAGO Dec. '08

BREEDING, LINE-BREEDING, IN-BREEDING, ETC.

The subject of breeding for best results in the poultry yard is exceedingly interesting, and is being developed more and more every year, not only by poultry breeders, but I believe by some of the government experiment stations.

There is "in-breeding," "line-breeding," "out-breeding," "cross-breeding," and no breeding at all.

Many people are afraid of in-breeding. By this is usually meant breeding brother and sister together for generations, without the infusion of new blood. This kind of in-breeding is very apt to result disastrously, because in such a flock the best, biggest and most vigorous are sent to the market, and the inferior ones are kept at home for breeders, unless a neighbor steps in and lends a cockerel to solve the difficulty.

For fear of the flock deteriorating, many people think it absolutely necessary to have new blood in their flock every year, and here is where the danger comes in for those who are raising thoroughbreds. If you buy pure-bred male of the same breed to mate with your pure-bred female from another strain or family, you may get one that will improve your flock, or one which will bring you disqualified birds. This getting new blood of the same family is called "out-breeding." J. H. Robinson says: "Most of the evils assigned to in-breeding are not due to in-breeding, but to careless selection. There is no evidence that in-breeding necessarily initiates degeneracy. There is abundant evidence that with proper selection for stamina to avoid common defects, very close in-breeding can be followed for a long time without injuring the stock. There is also abundant evidence that breeding unrelated fowls without careful attention to vigor, and avoidance of common defects is at once attended with precisely the same results as breeding fowls of near kin under the same conditions."

In making the new breeds, in-breeding is necessary to fix the color, shape, etc. If it is necessary to fix superiority in color, it is necessary to fix it in shape. If it is necessary to fix it in shape, it is necessary to fix superior laying capacity, for rapid growth and vigor. In-breeding is necessary because there cannot be intelligent breeding without in-breeding.

"Line-breeding," or breeding in line, is keeping to the same family, the same blood. It is very careful in-breeding. When we line breed we simply limit the number of ancestors in the fowl's pedigree. By so doing we intensify the qualities in the fowl, for it has been established beyond doubt that the mating of nearly related individuals has a tendency to intensify the traits or characteristics which they possess in common. As an example, I had a White Plymouth Rock hen (Snow Queen), a 95½ point bird. She laid 225 eggs in 9 months. I mated her, when I discovered her wonderful qualities, to my first prize male. Four of her daughters from that mating were prize-winners. The following year I mated her to her best son, and the third year to her son who was also her grandson. By this last mating, the offspring were 15-16 of her blood. I sold a few settings of this mating, one to a gentleman in Sacramento. He wrote me afterwards that he won first cock, first hen and first pen at the Poultry Show, with seven of her offspring; but, he added, "the great recommendation to your fowls is their wonderful vigor and healthfulness. All my other fowls have had roup and chicken-pox; in fact, I have lost more than half, and while yours were brought up with them, they seem absolutely immune to all sickness."

Another setting of eggs I sold to a party south of town. I heard later that one of the hens hatched from that setting laid 105 eggs in 110 consecutive days. By careful in-breeding it is possible to intensify the good qualities of great egg-laying and great vigor. A hen to be a great layer must have vigor.

To illustrate what is meant by line-breeding, I would take a good pair or trio of the best birds procurable; raise the young, carefully feeding for strength and vigor. The vigor of a flock is sustained not by introducing new blood, but by selecting breeding birds for vigor. Vigorous birds beget vigorous offspring; weak birds weak offspring, whether kin or not. The second year I would mate the father with two of his best daughters and the best son back to the mother hen, and use these two families as two different strains for new blood, each year selecting the best from either family. By the best, I do not mean the handsomest; I mean among the cockerels the most vigorous, active and up-to-standard birds, and among the pullets the best layers as well as the earliest maturing, largest and handsomest. Let it be understood that to breed from birds **because** they are related without making selections of points desired, is as wrong as to refuse to mate related fowls. By breeding from only vigorous stock, and observing the rule not to mate fowls having the same bad defects, mating together only fowls which in individual merit and in pedigree (whether akin or no kin) are what they should be for the purpose of the mating, you may be sure of avoiding mistakes.

"I am afraid of in-breeding," said a lady to me recently. "The book says change cockerels with your neighbor." I do not know from what book she was quoting, but I went to see her fowls. She had really fine standard-bred fowls to commence with, but she had ruined the flock by trading cockerels. A friend of mine intending to purchase them asked me to look at them, but I could not recommend them, as I knew the offspring would not be desirable.

Many persons wishing to purchase fowls from me (when I was in the business) would say, "Can you sell me two or four hens and a cockerel not related?" I replied that I could and would if they wished, as I had fifteen separate pens and marked all my young fowls, but if they asked me to mate for best results, I would give them hens from my best layers, mated to a cockerel that was partly related to them, for I knew then the offspring would be of as good quality as the parents. To know this takes some years of "close observation and close selection," which is the rule for line-breeding.

When I wanted new blood of late years, I would get a setting of eggs from the best breeder I knew. Select the two pullets from this brood, mate them with one of my own males, and then await results. Some years they would be quite satisfactory; if otherwise, they were consigned to the table and proved delicious eating. When the results were good, I had fine young ones and new blood which I knew would mate with mine and not deteriorate my fowls in regards to looks and standard points, but I could not tell for two years how the laying qualities of the offspring might be affected. Here is a place where "close observation" comes in. The pullets were trap-nested for a season, and then if they came up to my ideal I had the satisfaction of knowing I had made another success. This getting in new blood of the same breed is called "out-breeding."

I know a farmer's wife who had good pure-bred Plymouth Rocks, prize winners. She sent away and bought a first-prize winner—a beautiful cockerel. She thought she would have prize winners for the next show, when, to her grief, she found that all the progeny of that cockerel were disqualified birds. The cockerel did not "nick" with the hens, though they were of the same breed. This out-breeding was a failure. If she thought fresh blood necessary, she should have purchased a cockerel from the same breeder of whom she purchased her original stock, and she should have had one that had some of the same blood as the pullets, or if she could not do that, she should have bought a good pullet and mated her to the best male, and if the cockerel from that mating proved good she could have used one the following year. "Out-breeding" as she did, is a sort of lottery, and one cannot be certain of results.

Crossing, cross-breeding or out-crossing, all of which mean the same thing, is introducing blood from a distinctly different breed. The first cross will usually give better layers, and occasionally will produce good birds, but the progeny of these will be mongrels unless a pure-bred male is introduced each year. The new breeds, such as the Orpington, etc., are made by cross-breeding and then by close in-breeding. There is, however, one breed in America which has been made entirely by out-crossing; that is the Rhode Island Reds. This breed has been made by bringing vigorous blood on the male side "Red cocks" from China, Chittagong, Malay, etc., and mating them with the farm fowls of Rhode Island. This out-crossing has produced a breed of great vigor and prolificacy. Crossing, as a rule, is not advisable, because one can never be certain which parent the young will resemble; they will be large or small, some of one color, some of another, irregular in maturing and irregular in shape for market.

However, I knew a farmer's daughter in New York who wished to improve her flock of mongrels of all shapes and colors. She bought a "line-bred" Plymouth Rock cockerel, and the following summer she found that nearly all the young stock had Plymouth

Rock markings, even the offspring of the Cochin hens had feathers to their toes. The next year she bought again from the same breeder another vigorous Plymouth Rock, and by the end of that season she had, apparently, a flock of fine Plymouth Rocks. I say apparently, because if she had mated them together, she would have had mongrels the following season, but as it was she worked the mongrel old stock off and had fine looking Plymouth Rocks that proved excellent layers. A line-bred cockerel has greater prepotency than one indefinitely bred. That is, he will reproduce himself or leave his marks strongly upon his progeny. This was the case with my New York friend's birds. Hers were "cross-bred," or what farmers would call "grade" Plymouth Rocks.

The male bird, if he comes from a line-bred family, will be more prepotent than the female. He will impress his qualities or characteristics, good or bad, on his progeny more than a male that is not line-bred, and the male is considered half the pen. His part is the germ, the seed, from which will grow the chick. For this reason, choose the good, strong, vigorous cockerel, active and stirring, to head your pen and take a pure-bred instead of a mongrel, because in this way you will build up a flock of fine birds.

"Line-breeding" is keeping in the same family for years, each year choosing the most vigorous of both males and females to continue the succession. Line-breeding is very careful and closely selected in-breeding.

"Out-breeding" is introducing new blood, but of the same breed.

"Cross-breeding" or "out-crossing" is introducing distinctly new blood of an entirely different breed.

There is some diversion of opinion as to the best ages of parent stock to produce the strongest chicks, but it is usually accepted that fowls are generally at their best at twenty to twenty-four months of age. If they are not then in good condition, the breeder should look for something wrong in his method of handling stock. A hen coming two years old will, if properly handled between seasons, lay as well the second year as the first, and lay larger eggs which will hatch stronger and better chicks. A cock of the same age should be in his prime. The mating of males and females of this age will, other things being equal, give better results than any other age. However, well-grown young fowls would make better breeders than two-year-olds not in good condition. Many breeders advise mating a cock bird to pullets, and a cockerel to hens. Generally, these matings give better results than the matings of cockerels and pullets, but not as good as matings of two-year-olds.

The principal quality looked for in mating birds is vigor, whether you are mating for market or for egg laying or for fancy feathering.

Breeding Chart

A clear conception of the methods followed in line breeding may be had by reference to the accompanying chart which has been drawn from one published several years ago by I. K. Felch, the veteran Light Brahma breeder. In this chart the solid circles and segments represent the male blood elements, and the solid lines that a male has been chosen from the group from which they start. The white circles and segments represent the female blood elements and the dotted lines that the females have been chosen from the group from which they start. The shaded circle represents a scheme for the admission of new blood. Suppose we have two extra good birds which when mated together produce high-class offspring. Then the problem is how to perpetuate the quality of

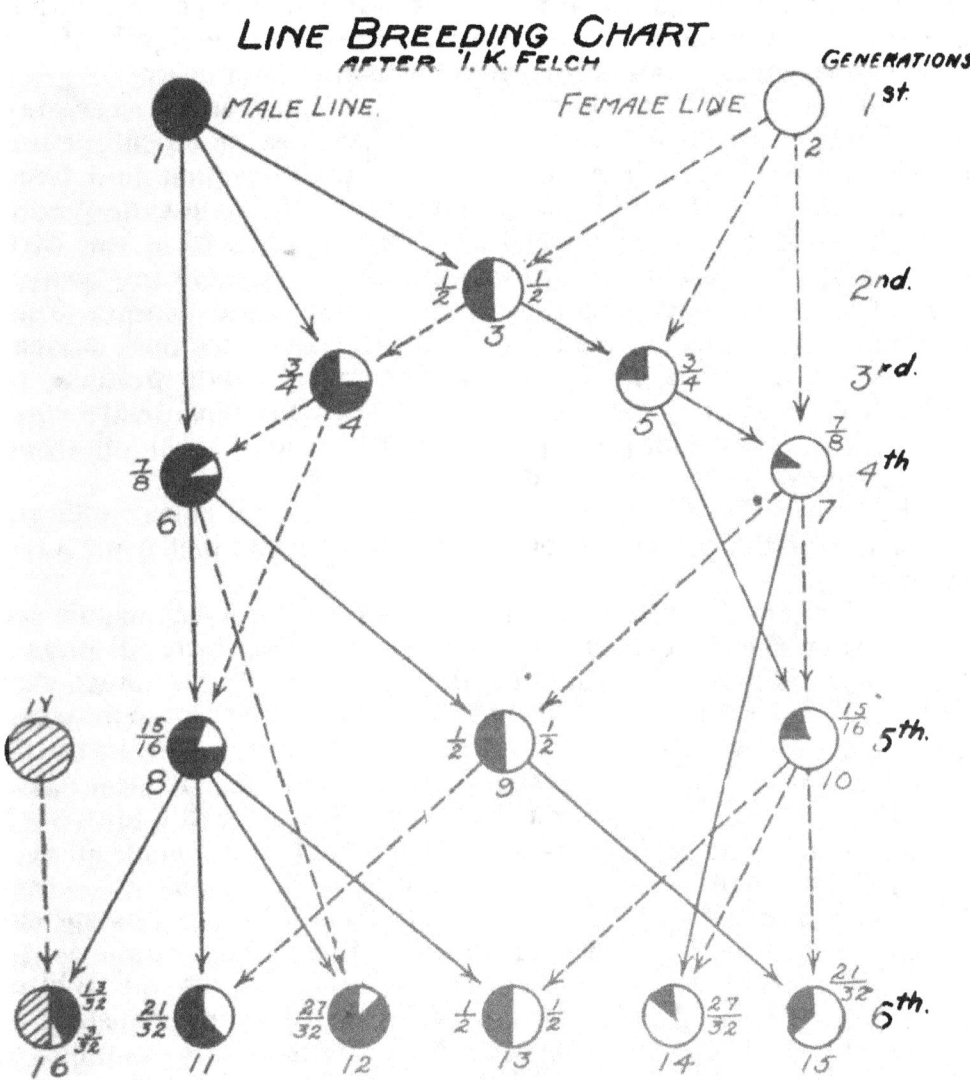

LINE BREEDING CHART
AFTER I. K. FELCH

By Courtesy of "Breeder's Gazette."

the parents and offspring without the dangers of close in-breeding or of destroying the results of several years of work, by violent out-crossing. By following line breeding, three blood lines may be developed, one of which shall contain a preponderance of original male blood, one a preponderance of original female blood, and the third equal proportions of original male and female blood.

In the chart let 1 represent the original male and 2 the original female. Then, by crossing 1 and 2, the result is group 3, which possesses equal parts of the blood of 1 and 2. Selecting the best pullet from 3 and mating to her sire 1, group 4 is produced, which contains three-fourths of the blood of the original sire and one-fourth of the blood of the original dam. In a like manner the best cockerel from 3 mated to his dam 2 produces group 5, which is made up of three-fourths of the blood of the original dam and one-fourth of the blood of the original sire. Proceeding in a similar manner by mating the original parents to their offspring in the third generation, we obtain at groups 6 and 7 offspring which contain either seven-eighths the blood of the original sire and one-eighth of the blood of the original dam, or seven-eighths the blood of the original dam and one-eighth the blood of the original sire, as the case may be. Thus the blood of the original sire has been practically eliminated from the female line, and the blood of the original dam from the male line. If the original parents were still in breeding condition, the blood of each could be intensified to 15-16 in the fifth generation. To obtain the original cross, however, at any generation after the second, it is only necessary to select parents from corresponding groups on each side of the line, as for instance, a cockerel from group 6 mated to pullets from group 7 will produce, in the fifth generation, group 9, which contains mathematically one-half the blood of the original pair. Similar results can be obtained by selecting parents from 4 and 5.

The fifth and sixth generations, as shown in the chart, indicate only a few of the possible groups that may be obtained from various matings.

The danger of using new stock not akin to one's own is far greater than the danger of line-breeding vigorous birds of known pedigree, as is proven by the following case in point: A breeder of Reds who had sold a fine pen two years previously of such excellence that she felt safe in buying back a cockerel raised from them, resembling in every way the sire sold, for the pen had been carefully line-bred and mated for best results; she used this bird with her finest hens and sold eggs from them, and every chick of this cockerel's get had white feathers enough to disqualify it. After this cockerel moulted, the beautiful red of his plumage was sprinkled all through with patches of white feathers. Close inquiry divulged the fact that the breeder had allowed his Reds and Whites to run together until a few weeks before mating time, trusting to this short period of time to remove the bad effects of commingling, Now, scientific men maintain that the danger of contamination of the female is far greater than is generally believed. Some even assert that a pullet's first mating influences her whole progeny, no matter how carefully she may be mated thereafter, and that the taint of foreign blood can never be eliminated from her offspring. There is much of truth in this theory, for the blood of the mother partakes of the blood of the sire through the blood of the unborn germ, whether egg or foetus, circulating through her. We can never be too careful to keep our hens and pullets safely yarded, and we should beware of strange males as of the plague itself.

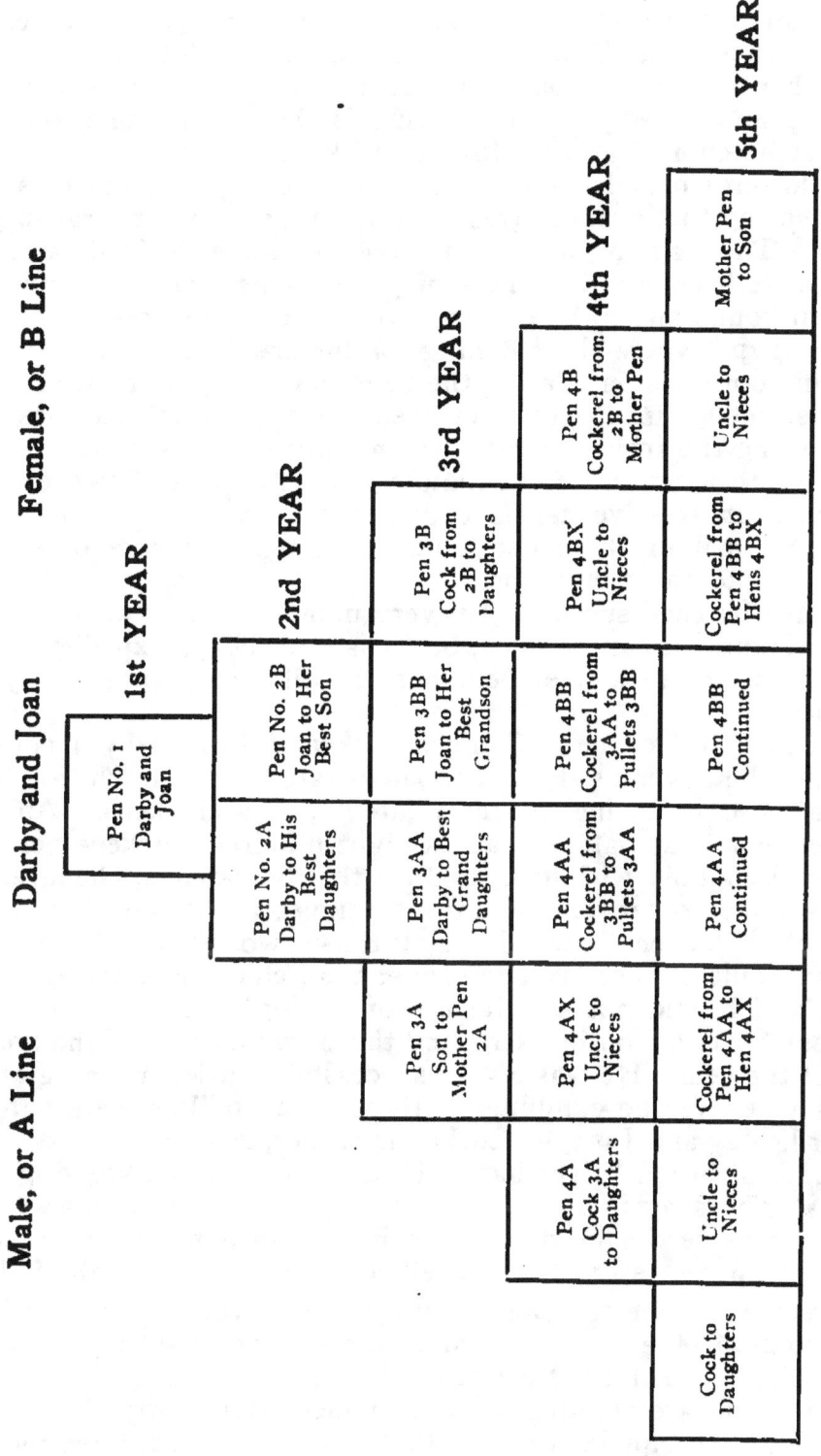

Male, or A Line   Darby and Joan   Female, or B Line

Mrs. Frank Metcalf's Breeders' Chart for Beginners

1st YEAR

Pen No. 1
Darby and
Joan

2nd YEAR

Pen No. 2A
Darby to His
Best
Daughters

Pen No. 2B
Joan to Her
Best Son

3rd YEAR

Pen 3A
Son to
Mother Pen
2A

Pen 3AA
Darby to Best
Grand
Daughters

Pen 3BB
Joan to Her
Best
Grandson

Pen 3B
Cock from
2B to
Daughters

4th YEAR

Pen 4A
Cock 3A
to Daughters

Pen 4AX
Uncle to
Nieces

Pen 4AA
Cockerel from
3BB to
Pullets 3AA

Pen 4BB
Cockerel from
3AA to
Pullets 3BB

Pen 4BX
Uncle to
Nieces

Pen 4B
Cockerel from
2B to
Mother Pen

5th YEAR

Cock to
Daughters

Uncle to
Nieces

Cockerel from
Pen 4AA to
Hen 4AX

Pen 4AA
Continued

Pen 4BB
Continued

Cockerel from
Pen 4BB to
Hens 4BX

Uncle to
Nieces

Mother Pen
to Son

WHEN TO HATCH

Solomon has said, "There is a time for everything under the sun," so there must be a best time for hatching chickens.

"When you are in doubt consult nature." If we consult nature she will undoubtedly say the spring is the proper time, the only time for hatching feathered fowl of all kinds.

Wild birds only lay their eggs in the springtime, and it is only in the spring that nature gives them the proper food for raising the young. Then, again, other birds, like the domestic fowls and pigeons, have been for thousands of generations trained to lay eggs the year round and to hatch them when man so desires.

In Egypt, where the hatching for thousands of years has been done in enormous hatcheries, the hens have lost the natural desire and instinct for incubation, yet even there the hatcheries are only used during the spring months, when nature would hatch.

The introduction of incubators into America, however, and American progressiveness have, one might say, changed the course of nature, making it almost a question of convenience or expediency when to hatch chickens.

I have hatched successfully every month in the year in California, so I know it can be done, but it is not always expedient to do so, and for the benefit of beginners I will relate some of my experiences.

When first I came to California I was informed by a poultryman that he never used his incubators later than March, for there was no use in trying to raise late-hatched chickens. Another breeder, who was hatching about eight hundred chickens per year and hatching only with hens, told me that he hatched whenever the hens wanted to set, except during August and September, as he found that chickens hatched during those two months did not mature any quicker, nor lay any sooner than chickens hatched in October, and he did not see the use of having to feed and care for chickens for two months extra, and those two months at the hottest time of the year. He was a very successful breeder, winning at the shows whenever he exhibited. His winning pullets were hatched in April, May and June, his cockerels earlier, usually.

The best brood I ever hatched came out on Thanksgiving day, the last Thursday in November. I found that the eggs as a rule were not as fertile in the fall as in the spring, and after some years of experience I refused to sell eggs for hatching in the fall.

I remember one time an eastern breeder, who had just arrived, came to me in the fall and wanted to buy several settings of eggs. I told him that I did not care to sell so early in the season, but he argued, "You are running your own incubators, why should you refuse me?" I explained to him that I could not guarantee the fertility of the eggs so early and rather than have customers disappointed I made a rule of not selling. I acknowledged that I hatched for myself at that season. He still persisted that nearly all the hens appeared to be through the moult, all were in vigorous health, were being fed the ration for fertility, and he would be perfectly satisfied with a forty per cent fertility, for he knew the difficulty of getting fertile eggs at that season. I let him have several settings and he was more than satisfied with the results.

I made a practice of hatching every egg that I could in the fall, commencing toward the end of September or first of October, and my reasons for so doing were, first, that it paid.

My object was to get cockerels that would be vigorous breeders at eighteen months of age or in the early spring, when they were about that age, while those that were culls would be in good condition either as broilers or fryers in the winter and bring high prices then, as there would be but few then on the market; so I was after cockerels first, and here I made some rather interesting discoveries which year after year were confirmed by a long series of experiments.

I had read (as I sometimes still read) in poultry magazines that the male birds ought to be entirely separated from the female during the summer, or at any rate during the moult, or whenever the eggs were not being used for hatching. I read that it was necessary to give the birds a complete rest, in order that they might be in a vigorous condition when they were wanted for breeding. Theoretically, this sounded plausible and I tried it, and to my great disappointment I found that the male birds that had been my best breeders did not regain their vigor for months, and some of them never again.

The organs of reproduction had become atrophied or dormant by disuse, and it was not until the following spring (nature's own season) that they became active again. Since making sure of this from many experiments, I have never entirely isolated the males that have been mated. I always leave two pullets with each. It really made me feel sorry for the male bird; he would at first fret so for the customary society and even when put with cockerels to police them and prevent them from quarreling, he seemed out of place and restless.

To have fertile eggs in the fall we must again consult nature and copy her rations at the time when she has fertile eggs—that is, in the springtime.

First—The rations then are tender greens just sprouting, and grubs and worms with but little grain, for then the grain is all sprouting. Then, again, by the springtime the fowls will not be so fat as they were in the fall, when nature provided a liberal supply of grains to give the necessary fat to keep them warm in the winter. By the springtime they will have lost that, still they will be vigorous and active. To have fertile eggs in the fall or at any time we must duplicate the conditions of springtime. Give an abundance of tender green, with sufficient animal food, and to induce exercise the grain should be fed in deep straw, so the hens will have to scratch and work for it.

When I began hatching in the fall my object was to get fall-hatched cockerels. It was cockerels I wanted. From past experience I had found that when I mated cockerels to old hens I usually had a large preponderance of cockerels, so I mated two-year-old hens to cockerels. Sometimes I mated only three hens to a cockerel, sometimes as many as ten. I also tried two-year-old males or even older with the hens in the fall, with the following results and discoveries, which I relate, not as conclusive evidence, for as the old saying is, "One swallow does not make a summer," and it would take more than one year's experiments at an agricultural station to prove that what I am relating as my experiences are always and conclusively the same.

First, I discovered that I could not absolutely govern the sex; secondly, I found that some males sired more males than females, and that it was not the most vigorous males that sired the most males; it was the smaller, younger or less matured birds; thirdly, I discovered, and this experience occurred over and over again, that the first eggs laid by hens after a period of rest, such as the moult, and also the first eggs of pullets, hatched more females than males and also that the darker shelled eggs (mine were White Plymouth Rocks) contained females oftener than males. I found that the first eggs laid after a rest were nearly always darker or else spotted with dark marks, and that when in the fall I was most anxious to get males, there were more females than males, as it was just after a period of rest. This was a disappointment to me at first, but later on I found that among those females hatched in October and November were my very best layers.

I might have known that pullets hatched from hens that were laying in the fall just when eggs cost the most would be likely to lay plentifully at that same time of year, but I did not realize it until the conviction was forced upon me by my own experience. Not all of the pullets were the best layers, but a preponderance of my best layers were hatched in the fall. After I had found that among my own White Plymouth Rocks the cockerels hatched in the fall were my best breeders and that the pullets were remarkably fine layers, I made it a rule to hatch every hatchable egg then. "Like begets like," if you hatch from the hens that are good fall and winter layers you will run a good chance of getting offspring from them that will come up to their mother's achievements, and if you know how to breed and mate them to males of the same good strain you will keep on improving your breed until you have grand layers.

A person can breed for whatever he or she wants—for fine feathers in the show room; for the best of table fowls for market; for prolific layers; for layers of large eggs, of colored eggs, or of white eggs. It is all a matter of selection and a study of the breeding qualities of your fowls. By just making up your mind what to do and how to do it and keeping the one object in view all the time you will succeed, there is no doubt about it.

But to return to our question: "When is the best time for hatching chickens?" Springtime, March, is the best time; but you can hatch also whenever your old hen wants to sit. You must remem-

ber, however, to give the chicks as nearly as possible the conditions
of spring. The reason that I say spring is the best time is because
it is nature's time and the eggs will be more fertile then; the fertility will be stronger and the chicks also. Very much depends
upon the way in which the parent birds are cared for and the food
they have. To have strong, healthy chicks the parents must be
vigorous; aim for vigor in your poultry yard first and foremost, and
then for whatever you most want.

GOODACRE RHODE ISLAND RED COCKEREL, PRIZE WINNER MANY TIME.

FERTILE EGGS

In the early spring we receive many letters of inquiry from beginners as to how they shall get fertile eggs or why the eggs are not fertile. It is of vital consequence to understand this matter somewhat.

To secure fertile eggs and strong chicks that will grow and make good breeders, that will be sturdy and vigorous and bring a profit, both the parents should be vigorous and healthy.

To grow vigorous chickens they must be well born and to accomplish this it is absolutely necessary to have the breeding stock in the very best of health. The females, as well as the males, should have entirely completed the moult.

The birds should be mature, both physically and sexually. This is a very important matter, for an egg may be fertile and yet not exclude the chick from it; the germ may not be vigorous enough to develop into a chicken capable of breaking its way out of the shell. More than mere size is needed in the male bird; maturity and vigor are necessary. The male should be of large and vigorous frame, well filled out, gallant to the females and ready to fight any intruder. He should have a full, deep voice and have lost the air of immaturity which the young birds always have. He should be ten months old or over, with hackles and sickles well developed and spurs of a fair size. Such a male will fertilize the eggs strongly and produce vigorous and sturdy chicks; the eggs will not only be fertile, but will be hatchable.

A male bird which is immature may fertilize many of the eggs, but it will be found that there are weak germs and many of these will never develop, or if they do, the chicks produced will be weak and inferior. Immature males are largely to blame for poor hatches and chicks dead in the shell. A cockerel is usually at his best when he is a year old, and from that time until he is three or four years old he can be used safely. During the breeding season the vigor of the male bird must be watched; he should have extra food with high protein content, that is, extra meat, to keep him vigorous. If mated to eight or ten vigorous females and if he is gallant, they will usually eat most of the animal food away from him, unless it is fed in the dry mash, and suddenly you may discover that the male that is heading your pen has lost strength and vitality with a corresponding loss in the hatchability of the eggs.

Much has been written on the importance of having fully matured and well developed females, but the best females cannot produce hatchable eggs if mated to an immature or weakly male.

I have found that two years of age is about the best for both sexes, otherwise have a year's difference in the ages of the pair of birds. Mate a one-year-old male to older females, say, two, three or four years older, or an older male to females of one year of age. Here in California I always try to have my male birds hatched in the fall; this was to make them at their best in the breeding season, fifteen or eighteen months later; also, I thought that males hatched in the fall would be the fathers of hens that would lay in the early winter, and I wanted fall and early winter eggs on account of the market price. I also found that my fall hatched pullets were earlier layers than the spring hatched; most of my record hens were hatched in November.

Another point in securing fertile eggs is to decide upon the number of hens that may be safely mated to a vigorous male. It has been found that the American breeds do best if one male is mated to from eight to ten females; with the Asiatic the number is one male to from six to eight females; while the Leghorns or Mediterraneans from twelve to twenty females, can be mated to a vigorous bird. These will strongly fertilize the eggs.

In my own yards I found that close observation was necessary; sometimes a male will apparently pay no attention to one or two females in his yard, and if after mating for three or four weeks I find the eggs from one of the females is not fertile, I remove her to another yard. I do not approve of changing the males in a yard, as some have advocated. The theory may be plausible, but in practice I have found it detrimental. It gives a feeling of unrest in the yards and retards egg production, as anything disturbing will, and causes a loss of fertility. I find it best to mate up for the season and then leave them alone without change of any kind, unless for some special cause.

A "line-bred" male is more prepotent than a male of no breeding, and will strongly impress his female offspring with the characteristics of the females in his line. Be very sure that your male is vigorous.

Feeding for fertility is another necessity in getting hatchable eggs. Here we may consult Nature. The spring is the time that Nature gives the fertile eggs; let us feed as much as possible as she does. Let us be sure to give plenty of tender, green, succulent food, as well as animal food to supply the place of the grubs and worms, which Nature gives, but in making any radical change in the food, make the change gradually. There is a difference between change and variety. A great variety will give fertility, while a radical change of any kind will cause a loss of eggs as well as loss of fertility. Among the green foods that give fertility, the foremost is alfalfa. Give the fowls all the alfalfa or clover that you can induce them to eat. Give all the grain in the scratching pen, so they will have to work and exercise for every grain. Nothing helps the fertility so much as the exercise of scratching and nothing costs much less. One prominent poultry breeder told me that it costs him exactly one cent a month per hen to keep fresh wheat straw in his pen; the hens scratch in that and also eat a considerable amount of the straw.

The grain most conducive to fertility is oats. I always use oatmeal in the dry mash during the breeding season, also sprouted oats. These are given besides the mixed grain in the scratching pen.

The animal food should be as much as possible, fresh green bone and meat, skim milk and beef scraps. The fresh meat is the best of all, but it must be fresh. Those living where rabbits or wild game are abundant can supply this. At the sea coast, fish and the little crabs or clams make a valuable addition to the animal food.

Of course, good, sharp grit, crushed oyster shell and charcoal should be before the hen all the time.

TESTING EGGS FOR INCUBATION

Success is what we all want to attain in whatever we undertake, and I earnestly hope that my practical talks on poultry may help others to make a success of it.

"Success with the Japanese," wrote George Kennan, in one of his interesting articles during the war, "is not a matter of perhaps or somehow or other, nor does it depend upon the grace of a merciful God. It is carefully 'pre-arranged' by an intelligent forethought, a perfect system and an attention to details that I have never seen surpassed."

Success in the poultry yard can be attained or "pre-arranged" in exactly the same manner. Failure in the chicken business (as in warfare) is due to lack of forethought, lack of system, and carelessness with regard to details. Forethought is the studying up and thinking how to do a thing, thinking out beforehand the best way of doing it and arranging for it.

The experiences of others by teaching us may save us not only dollars and cents, but chagrin and disappointment. I spend a good deal of my time in visiting the ranches of some of my correspondents, either to help them out of difficulties, or to mate up their pens for them, or to start up their incubators, or to overhaul their brooders or plan their henneries, and in this way I become acquainted with the needs and difficulties of a number of amateurs or beginners in the poultry business. Some of the troubles of others may teach us what "not to do."

"I wish you could tell me what is the matter," wrote one. "I had good luck last year, but only half the fertile eggs hatched last time."

I answered by spending a day at her ranch. "What is the matter with your hatches?" said I, "and on what day did they come out?"

"The first hatch this season came out on the twenty-second day," was her reply, "and as it was a day too late, I decided to run the machine half a degree higher than the directions order, and I suppose I got it too hot."

"Did you have any crippled chickens in the hatch?"

"Yes, in the last hatch there were a number of nice big chicks that could not stand up. Their legs sprawled out and I had to kill them."

The Incubator

Cripples usually come from overheating the incubator, or from irregularity of heat. Poor or insufficient ventilation will also cause cripples.

Now, what was the reason for these failures and what can others learn from them? After a careful examination of the incubator, which was a good one of the most approved make, I decided first that the incubator did not stand perfectly level; secondly, that the thermometer was at fault. When the incubator is in the least degree out of level, the heat will go to the highest side, leaving the lowest possibly a degree or more too cold. The first thing to be learned from this lady's failure is never to start the incubator without being absolutely certain that it is perfectly level. The only way to do this is to use a carpenter's spirit level. Put it on top of the machine at each side and then crosswise, and be sure that the bubble of air is at the proper spot. You may think that because it stood level last year it is most likely to be all right this year. That is leaving it to chance. One of the legs may have shrunk ever so little from the dry weather or swollen from the dampness of the room or the floor or ground may have changed ever so little at one corner or side without it being perceptible to the eye. It is much "better to be sure than sorry," so whether you are an expert or not, do not commence this season to hatch without testing your machine with a spirit level. Do not trust to luck—"pre-arrange" and success will be yours.

Test the Thermometer

Do not start the incubator this season without testing also the thermometer. Some friends of mine once bought a new incubator of standard make. The thermometer was guaranteed correct; two years seasoned. They had just received from Canada twenty dollars' worth of very choice eggs, and as they wanted to be sure of a good hatch from those prize eggs, they bought this new incubator, although they had a good one. Not an egg hatched! They afterwards discovered that the guaranteed thermometer was two degrees wrong. Do not trust to last year's testing. Thermometers vary, and it takes at least two years to season them.

It is not difficult to test a thermometer, but to do so you must have one perfectly correct and accurate. This you can either borrow from the doctor or from your druggist, or you can take one of your thermometers to the druggist and ask him to test it for you. Then, having one that is accurate, take a bucket holding about two quarts of water, put warm water heated to about 105 degrees into the bucket, and put your thermometers into it with the bulbs all at the same level. Keep the water well stirred, so the heat will be the same all over. Hold the thermometers in it for fifteen minutes, then read them and note the difference. If your thermometer is half a degree too low, mark on the incubator, "Thermometer half degree too low; run incubator half degree lower than directed," or opposite if the thermometer reads too high. If you buy a new thermometer, after testing it, be sure to hang or place it in the correct position. The bulb must be on exactly the same level as the former thermometer which belonged to the machine. A little difference in height or in the position of the bulb of the thermometer may make a great difference in the heat on the egg tray. You cannot be too careful and particular about these small items. "Pre-arrangement" of these means success.

How to Test the Eggs

After supper, when it was dark, we put the trays of beautiful fresh eggs on the dining room table, put the egg tester on the lamp, and then looked at each egg through the tester. Eggs were rejected that were chalky to the touch, or those that had light spots in them or freckled all over with clear places, or thin on the little end, or cracked, or crooked, or in any way misshaped. A few doubtful I left in, marking them "d" (these I subsequently heard did not hatch). It is much easier to detect the imperfect or unhatchable eggs by looking at them with the tester than by merely feeling them. It may be a little more trouble at the commencement, but is a saving in labor all through the period of incubation and a lessening in the expense of oil; besides giving more room for fertile eggs and more chance of a good hatch, as the infertile eggs chill their fertile neighbors and draw from their vitality. Therefore, do not put eggs into the incubator, or under hens, without carefully selecting them. Poultry keeping is made up of little things, and can so easily be ruined by little things that I will add a word of warning. Do not hold the egg when testing it so close to the lamp that it will heat it. The tiny germ of life in the egg is very tender and may easily be killed. For this reason I made a home-made tester out of a cracker box. I cut a hole the size of a half a dollar just opposite the place where the flame of the lamp came when I set it inside the box. In this way I did not overheat the egg. I also found this box very handy for testing eggs under setting hens. Eggs, whether under hens or in incubators, should always be tested out. There are thousands of eggs lost or wasted every year from carelessness in this matter. An egg which is infertile and is for a week either in an incubator or under a hen is perfectly good for food. It is simply an egg that has been in a warm place for a week. There is no germ in it; there never has been life in it, so there is no dead germ to decay. Infertile eggs keep fresh and sweet much longer than fertile eggs, and those who are raising only eggs for market should keep no male birds in their flock and never have fertile eggs.

NATURAL INCUBATION

The beginner may find it best to incubate with hens in preference to an incubator. The hen, having laid the egg, is the natural mother, has the mother instinct given by the Creator, and is certainly the one intended to hatch and brood the chickens. To the beginner in the chicken business there is less present outlay in a few sitting hens than in installing even a small incubating and brooding plant under artificial methods. The trials of those who find sitting hens troublesome are mostly due to their own inability or their lack of patience with the hen. Hens must be treated with patience and gentleness, for in no way can a hen that has the "setting fever," as our grandmothers called it, be coerced against her will.

How to Make Nests

The nest should be about fourteen inches square. Some breeders use boxes twelve by sixteen inches, but I prefer the square nests. If the nest is to be on an earth floor, rake the floor, then scoop a place about thirteen inches across in the form of a saucer; firm the shape well with the hand, and when it is smooth and firm, take hay or short straw, or tobacco stems, and firm that again in the proper shape, and the nest is made. Should it be necessary to have the nest in a box or on a board floor, take a clean box, have the front of the box just high enough to retain the nesting material; the backs and sides may be higher; put several inches of fresh earth into the box, firm it with the hand into a saucer-shaped hollow, and be sure to pack the earth high into the corners, so there will be no possibility of the eggs rolling into a corner and being chilled or lost. The nests should be flat at the bottom, shaped like a saucer and not like a bowl. If too deep, the eggs will roll together, sometimes pile up and get cracked or broken.

When only a few hens are to be set, the nests can be placed in any convenient location where the hens may be quiet, comfortable, away from other fowls and in the shade. I have found trap nests with two compartments very satisfactory, placed under a tree. I also have made sets of nests, giving each hen a nest and a small run, with a dish of water, a hopper with grit, corn and wheat always before her, shut off from all intruders. If hens are to be set in large numbers, a separate hennery in which from six to twenty hens can be set on the same day is the most convenient. The nests in this house or room should be placed with their backs to the wall and should face towards the center. Grit, corn, water and a dust bath for them to bathe in must be before them at all times. After a few days, if this hennery has a separate yard from the other fowls, the door of the house may be left open so the hens can go out of doors and take a dust bath in the open air, but the food, water and grit must be in the house in sight of all the hens.

Setting the Hen

The old-fashioned recipe was, "Set a hen between sunset and sunrise for luck." In other words, set a hen in the dark. Hens are quieter and not so easily frightened after dark. Choose quiet, gentle, tame hens; they make the best mothers. Handle them very gently. Put all the hens on the eggs in the same room the same evening, so they may all hatch out the same time. This is in order to keep the hens quiet during the hatch, as some whose eggs were not hatching the same day might become so excited they would leave their own nests and try to get to the newly hatched chicks when they heard the first peep.

Dummy eggs should be placed under the hens, when a number of hens are set in the same room, for a few days, a few under each hen. The first night after dark set all the hens on dummy eggs. If some light is necessary, turn the dark side of the lantern toward the hen. Have as dim a light as possible; move the hens gently, They will soon settle down on the eggs. In the morning look in and if any hen appears refractory, put her on the nest again and cover her with a box. Look in frequently for the first few days to see how they are doing, and you will rarely find more than two hens off and eating at the same time, as they are afraid of leaving their nests when others are off. Let the hens sit for two or three days, then put the good eggs gently in at night. The way to do this is to remove the hen gently, setting her on the floor; take out the dummy eggs and put the real eggs into the nest and gently replace the hen. Do not talk, act quickly, silently and swiftly, in a very dim light.

From thirteen to fifteen eggs are all that should be placed under a hen. It is all she can warm properly, all she can turn and attend to without the risk of breaking or cracking some. You will hatch more and stronger chicks by not placing too many under a hen.

Keeping Records

Above each nest, hanging on a nail, I place a card. On this card, legibly written is: (1) the date when set; (2) when due; (3) the hen's name or number; (4) name or parents' number on eggs; (5) number of eggs; (6) date of first test, number infertile or dead; (7) date of second test and remarks; (8) hatch, number taken from nest, number not hatching or killed; (9) toe marks of chicks. These cards can be preserved or copied into the diary of the ranch. They form a complete data of each hatch and a history of the hens as well as the chicks.

Testing the Eggs

Watch the hens rather closely for the first week, and note any that may be restless, nervous, cross to the others or stupid in not finding their way back to their own nests. These, when you test the eggs, you may be able to cull out and turn them back into the laying pen. It is always best to keep hens of pleasant disposition for mothers.

The eggs should be tested about the seventh day. An expert can test them earlier, and white eggs or duck eggs show the germ as early as the fourth or fifth day. The removal of the infertile eggs gives those that are left a better chance of hatching. The infertile eggs or dead germs are colder than the living eggs and chill the latter; besides, the infertile egg has a market value and can be used in the kitchen or fed to the chicks. It is a waste to throw them away. Testing should not be neglected. There is no use in hens sitting on eggs that will not hatch. They had better be reset on fresh eggs or returned to the laying pen.

Egg testers can be bought at the poultry supply houses, but a home-made egg tester I have used for years is only a box with the back knocked out and a hole in the top for ventilation. I put the lantern into it. Just opposite to the flame a hole about two inches square is cut in the box and a piece of a rubber boot-leg tacked on. I drew a pencil line around a fifty-cent piece and cut that out with a pen knife, leaving the round hole for the light to shine through.

The testing must be done in the dark. Set the egg tester with the lantern inside it on a box near the nest. Take the hen quietly off the nest, being careful to put your hands under her wings to make sure that you do not lift an egg or two with her. Place the hen very gently on the floor at one side. Do this so gently that the hen will not realize that she is off the nest. Take all the eggs from the nest, placing them either on the floor or in a basket; examine each egg and replace each fertile egg in the nest as you examine it; mark on the record card the number of infertile eggs, and gently replace the hen on the nest. Should any hen awake and appear nervous, she can be put upon the nest and the eggs slipped one at a time under her as they are tested, but the former plan is preferable, being more quickly done, with less disturbance to the hen.

The light shining through the egg, when held against the hole in the tester, shows the condition of the egg. Infertile eggs are clear. Fertile eggs have a shadow in them by the seventh day. The germ appears in some like a dark, irregular floating spot. Doubtful eggs should be marked with a D and given the benefit of the doubt, replacing them in the nest.

After taking out the infertile eggs, if there are many of them, you can reset the hens that have none or turn them back into the laying pen, culling out the fractious or nervous hens. By doing this carefully at each test, you will probably have good mothers when hatching time comes. Restless sitters usually make indifferent mothers. Close observation is necessary for success in all lines of poultry culture, and especially with setting hens.

The second test should be made in the same way on the fourteenth day. The eggs containing dead germs should be buried.

Dusting the Hen

A hen should be well dusted with insecticide the day she is set. To dust a hen the powder should be in a tin box with a perforated cover. An effective home-made peppering box can be made from a baking powder can with holes in the lid. Hold the hen by the legs, lay her on her side on a newspaper, raise the wing and sprinkle under it, then rub the powder well into the skin, especially around the vent. Work it into the soft feathers also around the neck. When one side is thoroughly powdered, turn the hen over and do the other side. The powder that is spilled on the paper can be returned to the can.

While the hens are on the nests they should be dusted on the seventh and fourteenth day and two days before the hatch comes off, with buhach or with any good insecticide. I prefer those principally made with tobacco dust.

When Hatching

In the climate of California I have never found it necessary to moisten hens' eggs. In fact, the eggs that contain dead chicks show that they have not dried out enough. They did not require more moisture. There is a natural perspiration which comes from the hen, and this keeps the eggs moist enough.

Should the eggs be chilled by the hen deserting the nest, do not throw them away. Put them under another hen as quickly as possible. I have known of eggs being left for a whole day and yet hatching. Eggs under hens will stand much more cooling than in an incubator. Chilling seems to be less injurious during the second week of incubation than at any other time.

On the nineteenth day, two days before the hatch, I take out to the nest a bucket of warm water, temperature 103 degrees; removing the hen from the nest, I put the eggs into the water. Those with a live chick in them immediately begin to bob or move as they float on the water, and I return them to the nest; those that sink to the bottom or remain perfectly quiet have dead chicks in them and will not hatch, and I mark them with a pencil; then replace the hen upon the damp eggs, feeling sure I will have a good hatch.

It is best to watch the hens pretty closely when the chicks are hatching. Some hens get excited and nervous when they hear the chicks peeping, and in their restlessness crush the shell so that the chicks cannot turn themselves and they die in the shell. These nervous hens should, if possible, be removed and quieter hens put on.

When chicks are hatching rapidly and the hens are nervous, it is best to remove the chicks as they dry off, taking them to the kitchen in a basket lined and covered with flannel. But if the hens are quiet it is best to leave the chicks with the mothers, only visiting the nests about twice during the hatch to take out the empty shells, lest they should slip over the yet unhatched eggs and so smother the chick. All eggs should be hatched by the end of the twenty-first day.

Marking Chicks

The offspring of the best, or pedigreed, stock can be marked so as to know them through life, by having a small hole punched in one or more of the webs of the feet. This should be done as the chicks are removed from the nests. A marker or punch is sold at poultry supply houses for marking chicks. They should be marked the day they are hatched, as the web is then soft, does not bleed as much as later, and there is not as much risk of the other chicks pecking the toes as they would do when older.

If the hens have been well cared for, properly dusted with a good insecticide during the three weeks of incubation, they will be perfectly free of lice. They and the chicks must be kept free. There is not the difficulty in this that many imagine. Dusting the chickens and hens once a week is all that is necessary. Some breeders put a little lard on the top of their heads and on their throats. This protects from the head lice. Others take a small brush (if the chicks are affected with head lice), and wash the little heads once a week with a lather of carbolic soap. They soon dry off in the sun or under the hen.

From experiments made in several stations it has been conclusively proved that hen-hatched chickens are stronger and heavier than those artificially hatched. At the Oregon Station the incubators hatched 78.5 per cent of "fertile" eggs and the hens hatched 96.5 per cent. The incubators showed 16.6 per cent of chicks "dead in the shell" and the hens 2.5 per cent. Chicks hatched under hens weighed heavier than chicks hatched in incubators, and hen-hatched chicks made greater gain in weight than incubator chicks, whether brooded by hens or brooders.

BLUE RIBBON WHITE LEGHORN HEN.

ARTIFICIAL INCUBATION

We are living in wonderful times, in the age of great inventions, and to succeed in any business, we must keep abreast if not ahead of our times. Not the least wonderful accomplishment of this wonder-working epoch has been the growth and advancement of the poultry industry, and the invention of the modern incubator, which made the development of the poultry business in this country possible.

In Egypt and China artificial incubation has been known and practiced for many centuries. In this country it is scarcely out of its infancy, still it would be impossible to estimate the value of the incubator to the poultry industry. It has made possible and profitable the large poultry plants in this country. It has developed the broiler business; it has raised the hen to the position of the money maker. One incubator will do the work of ten to thirty hens.

Must Approach Nature

There have been many kinds of incubators invented, made and patented in the last twenty years. The difficulty is to choose which kind will do the work of hatching eggs best; that is, will bring out strong chicks with the least attention and the least expense. There are hot water machines and hot air machines; round incubators and square incubators. I have heard of incubators in this state which are made like hot beds, heated with stable manure. Some incubators are heated with gas, some with electricity, but most of them by the heat of a lamp which burns coal oil. The best incubator is the one that comes nearest to imitating the natural process of incubation by a hen, for undoubtedly Nature is our great teacher in this matter.

The two favorite makes of incubators on the market now are the hot-water incubators and the incubators which bring warmed air into the egg chamber. The latter are called hot-air incubators. The difference between them is that the hot-water machines heat the egg chambers by radiation, while the hot-air machine brings warm air into the incubator.

In the machines where the heat is radiated from the metal surface of pipes or tanks, the temperature at the under side of the eggs, away from the heat, is several degrees cooler than at the upper side of the eggs. Top heat by radiation is supposed to resemble the heat from the body of the hen.

In the hot-air incubators the egg chamber is heated by air that is warmed outside of the egg chamber to a proper heat and is then forced into the machines by suction or circulation and diffused into the egg chamber. This way gives a constant supply of warmed fresh air, as pure and fresh as the atmosphere outside of the incubator. These hot-air machines rarely require any moisture to be added, as there is usually sufficient moisture held in suspension in the atmosphere, which is being constantly introduced into the egg chamber.

It pays to get the best, and by inquiring at the large poultry plants in the neighborhood, information can easily be obtained as to the most popular machine in use in that locality.

It is wiser to buy a machine than to attempt to make one. Good incubators are now sold at so low a price that it does not pay to risk the loss of eggs in experimenting on a home-made machine.

EGGS READY FOR HATCHING IN ROBERTSON'S HATCHERY. THE LARGEST IN RIVER-SIDE COUNTY, AT ARLINGTON.

Location of Incubator

The incubator should be located in a well-ventilated room or cellar that is dry and not subject to great variations of temperature.

Preparing to Hatch

The first thing to do is to set the machine perfectly level, using a spirit level to make sure of this, for if the machine is not level the heat will go to the higher side, the temperature will be uneven and although it may be correct where the thermometer hangs, in the middle, the upper side will be too hot and the lower too cold. It is most important to have the incubator stand perfectly level.

Let the incubator run for thirty-six hours before putting in the eggs. This is to make sure that the machine is thoroughly warmed and that it is running steadily at the proper heat. It may take twelve hours before the eggs gradually warm through, and the thermometer again shows the desired temperature. During this time the regulator must not be altered. Touching the screw may prove fatal to the whole hatch. So wait patiently until the desired heat is again present.

Selecting the Eggs

Eggs for hatching should always be carefully selected. The fresher they are the better. Eggs hatch after being kept a month, but the little germ or seed of life gradually grows weaker and weaker, and at last has not the strength to develop into a fine, healthy chick, and may die in the shell, if the egg is kept too long. Ten days or two weeks is better than any older.

The eggs should come from vigorous, healthy and well-fed stock. Much depends upon the feeding of the breeders, especially the male bird. They should have plenty of vegetables and green food, as well as animal food and those grains which contain the bone and muscle-forming elements. Eggs with imperfect shells should be rejected; also those with rough or chalky shells, and with thin spots· The eggs should be of medium size, neither too large nor too small, as the large eggs may have double yolks, which rarely hatch. Small eggs denote inferiority and are either pullet eggs or eggs from fat hens, or hens exhausted from having laid a long time.

Eggs of One Class

The eggs should be of one breed or class. It takes twenty-one days to hatch all hen eggs, but if the eggs from Leghorns are placed in the same tray as the Brahmas, the Leghorns will be the first hatched, sometimes as much as two days sooner, to the great detriment and loss of the others, which are slower in hatching. This is probably caused by the change in the atmosphere and temperature in the incubator at the time of hatching. The air is heavily charged with moisture, and the temperature always rises during a hatch from the activity of the chicks, and it is exceedingly difficult to regulate the temperature when the incubator is full of chicks in all stages of hatching. The rise of temperature does not hurt the chicks that are just breaking out of the shell, but if it takes place two days too soon, it will ruin the hatch of the heavier and slower breeds. Experiments that I have made along these lines have always given the same results.

Turning the Eggs

The eggs must be left for forty-eight hours after being placed in the incubator before being turned. After that they should be turned twice a day, or oftener. In this we should imitate the hen, for she not only turns her eggs constantly, but always shifts their position, pushing those that are on the outside into the center ot the nest. It is really more important that the eggs be moved or shifted from their position or location in the tray, than merely turned, as it shifts the location of the eggs in regard to weak germs or infertile eggs.

If the eggs are not turned during the early stages of incubation, many of the germs will dry fast to the shell and die, and the egg will be lost. When the egg is not turned during the latter part of incubation, the embryo does not develop properly, has little chance of hatching or may prove a cripple.

The turning and moving of the eggs gives exercise to the embryo; it is a species of gymnastics for strengthening the chick. The first forty-eight hours and the last forty-eight hours the eggs must not be turned.

Cooling the Eggs

Cooling the eggs I consider an important matter in our American incubators. The first week, following the hen's example, the eggs require but little cooling beyond the time it takes to turn them. The second week, as soon as the eggs are turned, replace them in the machine and leave the door open for five minutes; after this increase the time, a minute or two each day, till at the end the eggs are being aired or cooled fifteen or twenty minutes.

Cooling the eggs helps to make the shell brittle, so that the chick at the proper time can break its way out. Cooling the eggs contracts the shell and heating it up again expands it and this contraction and expansion gives the shell its proper brittleness. As the eggs warm up again, an almost imperceptible moisture comes over them, which takes the place of the perspiration of the hen, and obviates the necessity of sprinkling or dampening the eggs. So in our incubators it is necessary to cool the eggs. If this has been done properly the chicks will be strong and vigorous and few will die in the shell.

Testing the Eggs

All sterile eggs and dead germs should be tested out. Egg testers are sold with all incubators and very little practice will enable even a beginner to detect the sterile eggs and dead germs. Infertile eggs will be of a clear, uniform color throughout, except a slight darkening where the yolk lies. In the fertile eggs will be seen a small dark spot, and in a white egg the blood vessels can be seen branching out from it. Eggs should be tested about the seventh day. A second test for removing the dead germs should be made on the fifteenth day, they being easily detected at that time. The chicks in fertile eggs will be seen to fill the shell nearly, except a small space at the small end, and the air space at the large end. All eggs containing dead germs should be removed from the machine and buried. On the eighteenth day the chicks fill the entire shell except the air cell, and the egg will be quite opaque, as if nearly full of ink. To become accurate in egg testing requires practice and a brilliant light.

Operating the Incubator

Follow exactly the directions given with whatever incubator you may purchase. The makers of the incubators are anxious for you to succeed and have good hatches; it is to their interest for you to be successful. They have spent time and money in perfecting and understand how to manage their own machines better than anyone else.

On the morning of the nineteenth day the eggs should be turned for the last time. The machine should then be closed and kept closed until the hatch is over. Opening the door during the process of hatching may spoil or seriously injure the hatch, as by such action a large amount of heat and moisture escapes and cold air is admitted. This dries up the lining skin of the eggs that are pipped and checks or prevents their hatching. It also chills the half-hatched or newly hatched chicks and is detrimental to all of them. When the chicks are coming out lively, the temperature will rise: should it go above 105 degrees, the lamp may be turned down a little.

Leave the chicks in the machine without opening it until they are thoroughly dry. The chicks should not be moved from the incubator until the twenty-second day and should not be fed until thirty-six hours after hatching.

General Remarks

Should the hatch not come off until after the twenty-first day, it shows that the heat has been insufficient; if it comes off earlier, the heat during part of the time has been too high. Too low a temperature will give a weak hatch, many chickens will die in the shell, and those that are hatched will be weakly and never amount to anything. Too high temperature at the commencement of incubation will cook and kill the germ. One hundred and six degrees is danger point up to the tenth day. Germs which died between the first and second testing are frequently the result of overheating. Too high a temperature during the last week will so weaken the bowels of the chicks that they will be unable to assimilate the yolk of the egg. The yolk of the egg is Nature's perfect nourishment, which feeds and nourishes the embryo.

During the last day of the chick's life in the shell the part of the yolk which has not been absorbed is drawn up into the chick. This forms its food and nourishment for about three days. But should the egg be overheated, this yolk hardens and even if drawn into the chick, it becomes tough, the chicken's bowels are weakened by the overheating, the yolk remains unassimilated, like a piece of rubber, blood poisoning ensues and the chick dies some time between the first and tenth day of its life. Chilling the eggs has almost the same effect; it weakens the bowels, hardens the yolk and eventually kills the chick.

The incubator is a splendid hatcher of all kinds of germs, and white diarrhoea may be caught in the incubator. Infection may be conveyed through the shell of an egg or even exist in it before it is laid, thus carrying to the embryo chick the germs that lead to its early death. To prevent this some of the latest investigators thoroughly disinfect the eggs and the incubator before the hatching is begun. To accomplish this the eggs are wiped with a cloth dampened with alcohol, and the incubator is washed with a solution of some antiseptic, such as creolin, in every part of the inside, and the egg trays washed and then set in the sun to dry and air.

CARE OF BROODER CHICKS

The hatching of chicks is but half the battle, for eggs from good, vigorous parents will hatch with but little trouble if a good standard incubator is used and if the directions with it are followed. How about the raising of the chicks after they are hatched?

The poultry papers agree that there is not a subject pertaining to poultry culture that needs more thorough, painstaking investigation and discussion than the care of the chicks, and it is said that not more than fifty per cent of the chicks that are hatched the country over reach maturity or a marketable age.

What are the principal causes of mortality among chicks; how can we combat them and what are the essentials in the successful raising of chicks?

BROODER HOUSES, 14x20, DAVISON RANCH, ARLINGTON. EACH HOUSE BUILT TO CONTAIN 1500 CHICKS. WITH AUTOMATIC HEATER.

There are numberless causes for the death we deplore—among these are diarrhoea, bowel trouble, lice, improper feeding, impure water, overheating or chilling and exposure to the elements.

Feeling sure that the mortality in chicks is caused in a majority of cases by the carelessness or ignorance of the caretaker, let us discuss this subject and glean from the best authorities some ideas about it as far as we may in one short article.

Expert Opinion

Prof. James E. Rice, of Cornell University, has for several years been making a careful study of the cause and cure—or prevention —of the numerous diseases that cause the death of hundreds of thousands of chicks yearly, and his investigations have led him to believe that one great cause of mortality is the failure on the part of the digestive organs of the chicks to properly digest the yolk of the egg remaining in their bodies at the time of hatching.

Mr. Rice says: "If we can solve this one problem—the cause of the anaemic condition of chicks that follows this failure to absorb the yolk of the egg—more money will be saved in one year to the farmers and poultry raisers of New York state than it costs to run the State Agricultural College for ten years."

Mr. Rice says he is confident that environment has little, if anything, to do with the disease, as has been generally supposed. When he first began his investigations, this theory was worked upon and followed up, but as the investigation progressed it was found that the same conditions existed under almost any and all circumstances—in dry places, in damp places, in light brooding houses and in dark brooding houses; in fact, he found no conditions under which this trouble did not exist. Mr. Rice is confident, however, that the investigations being conducted will ultimately solve the problem.

Until this problem is solved we shall have to be content with the theories of the different breeders and hatchers, and as one I feel confident from my own experiments and experiences that the deaths from diarrhoea, or in fact almost all the deaths of brooder chicks before three weeks of age, come from faulty incubation. The temperature has been either too hot or too cold, usually the former, or the ventilation has been at fault, or the chicks have been chilled in carrying them to the brooder, or fed too soon, before the digestive organs were ready to digest the food.

Elbow Room Needed

Mr. Hunter, the veteran poultry man, says: "With incubator chicks raised in brooders, elbow room seems to be a most important factor, and want of elbow room is one cause for the great mortality in brooder chicks."

It is quite natural to suppose that a brooder which is three feet square is abundant room for seventy-five or a hundred chicks, and indeed it is for the chicks as they come out of the incubator, and if we do not want them to grow it might be all right to crowd them into the brooder, but these chicks will be almost twice as large at three weeks old as when they are hatched and will require twice as much room or will suffer for it.

Fifty chickens are as many as should be put into any brooder. To increase the number beyond that point will induce crowding, which kills some and stunts others, and will prevent the quick, healthy growth that is necessary for all young animals. Ample brooder room is the first and chief requisite for the health and comfort of the chicks. The next requisite is oxygen. In other words, plenty of fresh, warm air, but no drafts in the brooder. Here is one of the great faults with many brooders, as for example the hot water pipe brooders in use in many brooder houses. Those hot water pipes merely heat the air that is already within the hovers, which air is practically confined to the hovers by the felt curtain in front, provided to keep in the heat. It does that, but it also encloses the air, which the chicks have to breathe over and over again.

This defect in my brooders cost me the lives of many chicks before I discovered the cause. A current of warmed fresh air supplied under the hovers overcame this difficulty, when I submitted the hot-air plan.

Comfort Essential

The brooder should be heated for at least twelve hours before the chicks are put into it. I always keep a thermometer in the brooder and have it at 95 degrees when they are first removed from the incubator. They should be carried to the brooder in a basket lined and covered with flannel, great care being taken that they be not chilled on the way. I am sure that many chicks lose their lives by being chilled on this their first journey. The abrupt change from the warm incubator to the outside air, which is thirty or forty degrees colder, is sufficient to chill the chick.

NIGHT SCENE IN 14x20 BROODER HOUSE, CONTAINING 1300 CHICKS, DAVISON RANCH, RIVERSIDE.

A chill will harden the yolk of the egg, which is drawn up into the chick the last day of its stay in the egg shell. You know that the yolk of the egg forms the nourishment for the chick inside the shell. The last day of its life in the shell all that remains of the yolk, about one-fourth of it, is drawn up into the chicken through the navel. If the chick is vigorous the yolk should be assimilated or digested in about three days. But if the chick is chilled or overheated, it so weakens the bowels that they cannot digest the yolk or absorb it, and the yolk hardens or toughens, becomes almost like rubber; then it can never be assimilated, blood poisoning ensues and the chick's life ends.

Chicks should not be fed for from thirty-six to forty-eight hours after they come out of the shell, because, first, they do not require any food, as the yolk inside them takes nearly three days to become absorbed or digested; and, secondly, if they are fed too soon (that is, before the yolk is digested), the effort of digesting the new food draws the nervous energy or gastric juices away from the part containing the yolk, up to the crop and gizzard, and the yolk either does not digest at all or digests so slowly that it brings on bowel trouble, which at such an early age stunts the growth, if it does not kill the chick. In a chick that is fed too early in life the yolk will take, or may take, ten days to digest. You ask how I know this. "By sad experience and post mortem examinations," is my reply.

The brooder being warmed to a temperature of 95 degrees under the hover, the floor should be covered with coarse, sharp sand, the chicks carried carefully to the brooder, after remaining thirty-six to forty-eight hours in the incubator.

Feed Carefully

The first few hours in the brooder they require no food but the sand to eat and water to drink. The sand supplies the little gizzards with the necessary teeth or little grindstones, so that they are ready to commence work when the food comes. Water I place in a drinking fountain, so they cannot get into it and wet themselves. I give them water from the first. I know some people do not, but it has succeeded well with my chicks. At about four o'clock they have the first meal. I scatter rolled breakfast oats on the sand. The white flakes quickly attract their attention and they pick them up. I also give them a fountain of fresh water and one of sweet skimmed milk. It is surprising to see how quickly they learn to eat and drink. In the evening I look in upon them and am pleased when I see them spread over the hover floor, as it indicates that they are comfortably warm and will not crowd or huddle during the night. The first thing in the morning I give them some more rolled oats and some "chick feed." The "chick feed" I buy at the poultry supply stores. It is composed of a variety of seeds or grains, with a little charcoal, dried blood, or beef scraps and grit. Sometimes I make my own chick feed by mixing cracked wheat, kaffir corn, millet, steel cut oats, pearl barley and rolled oats together, adding charcoal and dried beef scraps. I put more wheat and more oats into this mixture than any of the other grains. The chick feed that I buy has in addition some other seeds, such as rape or mustard, canary seed, hemp, etc. I buy chick feed to save myself the trouble of mixing. Chick feed and rolled oats is their main feed until they are six or eight weeks of age. I feed them five times a day at first, and I always leave a little feed trough or hopper of chick feed where they can get it. I know this is contrary to the advice of many, but I found the weaker ones did not get the proper amount when all rushed for the food, and also it was a great comfort to me, if anything detained me beyond the usual feeding time, to know they had food before them. Also when fed at the usual hour they were not so ravenously hungry; they would not overload their little stomachs.

Their morning meal at about six in the morning, consists of rolled or flake breakfast oats, next green feed, then chick feed, then rolled oats, green feed and the last feed after they are a few days old is hard boiled eggs (two for every fifty chicks), chopped fine, shell and all, mixed with dry bread crumbs or cracker crumbs, and an onion chopped very fine. I mix all together, adding a little pepper and salt. If I have no bread crumbs, I add Johnny cake or rolled oats to the onion and egg. I always send them to bed with their little crops full.

As They Grow Older

I keep a thermometer under the hover in the brooder and lower the temperature one degree a day until it is down to sixty-five degrees. After the chicks are six weeks old, unless the weather is unusually cold, they require no heat. For green feed they seem to prefer lettuce to anything else. Finely cut clover or alfalfa is excellent. The lettuce I cut up very fine at first, but in a few days they learn to tear it up, and lettuce suspended on a string or even thrown on the ground, gives them exercise and amusement as well as food.

In the playroom, where the chicks are fed, the floor is covered with chaff. If I cannot get from the mill real chaff I cut up hay in the clover cutter, either wheat hay or alfalfa hay, to give them something to scratch in, and I throw a handful of chick feed into it for them to have something to reward their efforts.

The alfalfa hay or chaff keeps them busy and exercising and this broadens their backs and increases the size and vigor of the egg making organs which are already commencing to grow and which we must develop from the very first if we want to increase the egg output. The chaff, or preferably the alfalfa hay chopped short, also conceals their little feet from their active and sometimes mischievous brothers and stops them from pecking the feet and drawing blood, which tastes so good that they will actually turn cannibal and tear out and eat the bowels, sometimes causing great loss. This is always prevented by keeping the chicks busy scratching in deep chaff.

They have fresh water each time they are fed. The first meal is at about six in the morning, and if I fear that I may be later than that, I put fresh feed and water in their playroom over night, so that the hungry babies may not be kept waiting. They come out at daybreak, eat a little, and sometimes drink, and then go back and take another nap.

The brooders must be cleaned twice a week the first week, three time a week afterwards, and every day when the chicks grow larger. The chicks should be dusted with insect powder about once a week. To do this I have a tin box (a baking powder can with a perforated cover), put insect powder into it and after dark raise the hover and sprinkle the powder liberally over the chicks. This will usually keep them free from lice.

FIRELESS BROODERS HAVE COME TO STAY

Fireless brooders have come to stay, at least in California. I do not mean to say that they would be suitable in a broiler plant, for there chicks are raised not to be muscular and sturdy, but tender and fat, and for that they require to be kept always warm and fed a fattening diet, and the heated brooder is or may be better adapted to their needs, but for the sturdy chick, the chick we want to develop into a first-rate layer, or a large market fowl, or a winner at the show, the fireless brooder, properly handled, in this climate is excellent.

Some few months ago I gave a description of a home-made fireless brooder which one of our readers made two or three years ago. Several made some by that plan and have expressed their great satisfaction at the ease with which they now raise their chickens. At the same time I mentioned that many of the poultry supply houses had excellent fireless brooders for sale. Since that time I have met a number of prominent poultry breeders here, who had been quite prejudiced against these fireless brooders, just as many poultry raisers years ago thoroughly disapproved of incubators, and I find those who have tried the brooders without heat are loud in praise of them.

One very successful business man who wins prizes every time he exhibits, said to me: "These fireless brooders are great. I have not lost more than three per cent of my hatches since I have used them." And in talking over the brooders with many others I find that one of the great advantages is that there is no fear of fire. Where no fire is, there is no danger of either smoke or a conflagration, which is a very great comfort to a busy poultry man or woman, and especially at night.

I have lately seen a brooder made by Mr. Hammons, the manager of the mammoth broiler plant near Los Angeles. It is easily made and has some points of special value.

The brooder made by Mr. Hammons is his own invention and he has no objection to any one copying it. It is a box 20 inches square and 6 inches deep, and in each corner has a small block 4 inches high for the frame of the hover to rest upon. The lower frame does not fit tightly in the box; this is one of the new improvements; there is a space of about a quarter of an inch on all four sides; this is for ventilation. A door four inches square is cut

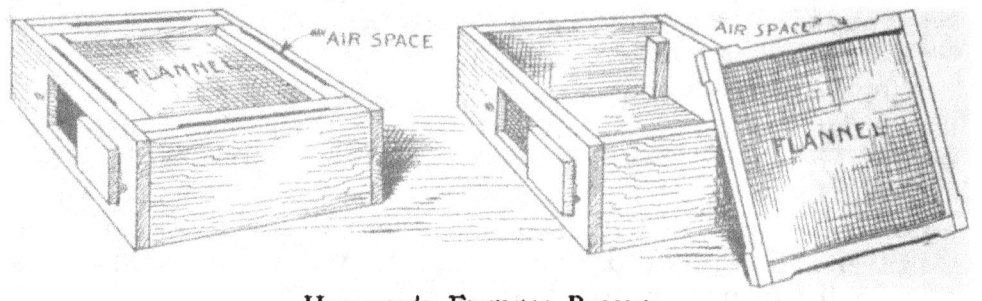

HAMMON'S FIRELESS BROODER.

and hinged on one side of the box for ingress and egress of the chickens. The hover frame is covered with heavy double canton flannel, and seven square blankets cut out of good thick felt lie on top of the hover. These little blankets must not cover the quarter inch crack for ventilation, but should just fit inside the frame. This is another special novelty. The first week all of the blankets are used and each succeeding week one is removed, until at eight weeks of age the chicks have no blankets over them and are ready to leave the brooder.

The brooder 20 inches square and made as I described will accommodate only 25 chickens. Mr. Hammons' experience has taught him that this number is the very best for one flock, as then each chick can grow without crowding.

At first he makes a nest of straw nearly filling the box, leaving a nicely rounded out place in the middle for the baby chicks to nestle in, and as they grow, less straw is needed, but a little should always be used to keep the floor and the chickens' feet clean. The blankets should be sunned and aired daily to keep them sweet and clean, as one airs one's own bed.

Mrs. Frank Metcalf, the originator of the celebrated "Buckeyes," writes: "I have had fine success with Mr. Hammons' brooder and recommend it to others as the best I have ever used. I raised forty-seven out of fifty hatched in the last batch of Buckeyes. Fifteen turkeys may be raised in one of these; I found that eleven did very nicely, although more would have been better at first. We had little coops 30 inches wide, by six feet long and confined the chicks with the box inside of these for the first week; after that they had wire runs out of doors."

This brooder is simply a square box, 20 x 20 inches, 6 inches deep, made of ¾-inch dressed tongue-and-grooved wood, with a hover laid on it instead of a lid, and with ventilation all round the edge of the hover and the sides of the box, giving free air around the chicks as it would be around a hen. It is a good imitation of a hen.

Handles can be nailed on the box so it can be carried easily, chicks and all.

The canton flannel cover of the hover should have a little plait at each corner, so the flannel will sag down in the middle a little, on the backs of the chicks.

I have found that if too few chicks are in the brooder that they cannot at first keep warm enough. Six chicks for instance are too few. In that case I put a hot water bottle or bag on the top of the hover, under the blankets, for a short time. I also have found that the blankets can be cut out of a common woolen blanket, which does as well as the felt and costs less.

"WHITE DIARRHOEA" IN BROODER CHICKS

This is a disease which rarely attacks chickens hatched and raised by hens, and therefore it must be caused either by faulty incubators or wrong "mothering."

We all know that at times quite a number of chicks in a brooder will be "stuck up behind," as it is sometimes called; how they run about with their shoulders up, looking wizened and old; how they try to huddle near the warmth and finally give up the hopeless struggle and die.

"I think my chicks are taking some disease and dying from an epidemic," said a lady, who, though a novice with incubators and brooders, was an old and most successful poultry woman with hens. These chicks had been overheated in the incubator I discovered two days after hatching.

Another friend, a very clever surgeon, told me one chilly night his incubator lamp went out and all the eggs got stone cold. His wife could not bear to think of losing all those nice eggs after having watched them for nearly three weeks, so she advised lighting up again in hopes of saving some. This they did, and were rewarded with fifty nice, lively chicks, but in a few days they commenced to die; they were "stuck up behind," or they shivered and seemed quite thirsty, and at last, when only fifteen were left, he made some post mortem examinations, and he found the yolk of the egg, which is drawn up into the bowel cavity the last day of incubation, was still there, only it looked in some like a bit of rubber, in some like hard-boiled eggs, and again in others it was dark and putrid. Instantly he reasoned that it was that yolk that was killing the chicks by blood poisoning.

He had only fifteen left, but he decided to experiment on them, so he opened them; his wife begged him to give them chloroform, which I believe he did, and he removed the toughened yolk, sewed up the wound, fed them lightly and all of the patients recovered and lived to maturity.

It was a delicate operation, but my friend had the skillful hand of a trained surgeon. I never attempted it myself, but have made many a sad post mortem on little chicks dying from being "stuck up behind," for I make it a rule to hold "post mortems" on all subjects that die in my yards.

One time a whole incubator of eggs—240—were overheated by a meddlesome child playing with the regulator. Two days later 117 hatched, the others were cooked hard. Every one of the 117 died, although some lived to be eleven days old. I did everything I could think of to save them (except the surgical operation), but lost all.

I feel sure that either overheating or chilling so weakens the bowels that they cannot digest, or, rather, assimilate the egg, and that the yolk putrifies and causes blood poisoning; and that either overheating in the brooder or chilling before the chicks are a week old will have the same result. Also if the chicks are fed too soon after hatching, the digestive juice or whatever it may be called, goes into the crop and gizzard to digest the new food and the yolk of egg is left either to digest very slowly or not to digest at all. In either case it will give diarrhoea and it may end fatally.

I am often asked what to do for young chickens that have diarrhoea, and also for those that are "stuck up behind." I know how almost hopeless these cases are, as they usually come from the unassimilated yolk of egg, but I reply that rice boiled in milk, adding a tablespoonful of ground cinnamon to every pint of milk is about the best remedy for diarrhoea that I have tried, and to pick off with the fingers the dried excrement, slightly greasing the vent with carbolated vaseline is the only way for "stuck up." If the droppings are washed off, it is almost sure to chill the already weakened bowels and result fatally.

Incubators, To Disinfect

Dr. Woods recommends that incubators be thoroughly scrubbed with a solution of one gill of creolin in 8½ quarts of water each time before putting eggs in them, to prevent the chicks from contracting white diarrhoea and other bowel troubles. The machine should be thoroughly dried before putting the eggs in. Every part should be scrubbed inside and out, and the egg trays should be especially well done. If the eggs are also disinfected there is very good reason to believe that the ravages of white diarrhoea will be largely diminished.

"TOPEKA," MUMFORD & EMERSON'S FIRST PRIZE WHITE PLYMOUTH ROCK.

VIGOR

I never advise beginners to commence by trying to make a new breed, because very few are capable of success, just as there are but few artists who can paint a magnificent picture when they first begin to paint. To beginners I say, choose the breed and the standard that you like best, and keep to that breed. Then go on improving your flock. The way to do this is first of all, look to the vigor of your flock. It is VIGOR, first, last, and always that you want. "But," says the beginner, "how am I to get vigor, and how am I to keep it?"

First to get vigor, you have to begin with the parents.

Get your eggs from healthy, vigorous stock, that have been fed the ration for vigor. Then hatch them properly, remembering that if you have a poor hatch (that is to say, if you find a number of chicks dead in the shell, if the hatch has been hurried by too much heat or retarded by too low a temperature), that those chicks which do manage to get out of the shell will not have vigor of constitution, nor size of frame, nor the early development so necessary for success. A great deal depends upon the chick being properly hatched; for that reason I advise beginners to commence hatching with hens, and when they do have an incubator, get a good standard incubator, and set one or two hens at the same time, keep them both running evenly together. Biddy will teach beginners a great deal. Then when the chicks are hatched, feed for vigor. Consult Nature, feed the fluffy little fellows after you have allowed them the necessary rest of at least thirty-six hours before feeding them. All a chick needs is rest and warmth to go on growing for about two days or even three; after that time its digestive organs are ready for work; then they must have the proper kind of food.

The Crop

Nature has given the chick a crop where the food is first received. In this crop is found a fluid, something like the saliva in human beings; this saliva acts upon the food, softening it and otherwise preparing it for digestion. The food then moves on to the proventriculus, or stomach, where it is still acted upon by a fluid, and it finally passes to the gizzard.

The dry chick feed, so universally used, composed of a great many fine grains, is admirably adapted to feeding the chick. There are some grains especially conducive to vigor; the chief of these is oats, in any form, steel-cut, hulled, or rolled breakfast oats. There is another thing which Nature in the spring time gives the chicks, plenty of worms, bugs, insects. Often after an April shower I have seen the ground covered with worms, but here in California there are not enough insects to supply the chickens, therefore the chicks must have animal food as well as succulent green food. I used to buy two pounds of hamburger steak three times a week, and nothing suited the chicks better, fed raw once a day.

Exercise

Vigor comes from exercise as well as from the proper food. Scratching is by far the best exercise for chicks. It keeps the organs of digestion in a healthy condition; it gives the chick a good appetite; it broadens the back, giving plenty of room for and developing the egg organs, strengthens the muscles and enlarges the frame.

How shall we give them work? The best way, of course, is to give the mother hen range. Chicks on range with the mother hen rarely acquire bad habits. It is chicks in the brooder that get into mischief, that quarrel and scrap, peck each other's toes and get to be cannibals. The best way of preventing mischief is by bedding the brooders, one or two inches deep, with alfalfa hay, cutting to half-inch lengths in a clover cutter. The little chicks will eat some of this, and they will scratch in it for seed of the chick feed all day long. This chaff, or finely cut hay, hides the toes so they will not be tempted to peck each other's toes. Another method for exercise is planting the runs with wheat or barley. The chicks will scratch up or pull up the green sprouts. Hanging a head of lettuce up in the brooder house will also afford both amusement and exercise.

Never let chicks be crowded at night. Many a chick that might have been a prize winner is disqualified, has off-colored feathers simply from having been crowded or bruised by a larger chick treading on it. A bruise, even a slight one, will often result in a white feather on a colored fowl or a black or red feather on a white fowl, and overcrowding has the same effect.

More About Vigor

Vigor has always been one of my "hobbies." I have written much about it but must add a little more. Breeding for vigor is one of the problems most interesting in poultrydom. It might not be difficult if we could closely imitate Nature, but we are demanding much more or our hens than Nature does, and here is the point where we fail.

Much of the lack of vigor, the low hatches, weakness and mortality of chicks, and the inferiority of the mature fowls may be traced to the so-called "intensive" methods and of forcing the hens to produce an abnormal number of eggs with a consequent breaking down of the constitution of the hens. The intensive system of keeping fowls in small quarters and feeding them with stimulating rations has contributed largely to the lowering of constitutional vigor in many large flocks of hens.

There are in nearly all flocks hens that differ in vigor. There are weak fowls and strong fowls in all the different breeds and what we want to aim for is the strong, vigorous hen that will digest the most food and lay the most eggs. Let us study how to get these hens. First, cull closely; that is, get rid of, market or eat those fowls that do not come up to the mark in vigor. Secondly, let us mate together only those that have constitutional vigor. It is often a temptation to breed from a hen that may have won a prize, or have beautiful feathers, or is high scoring, but has had roup or is defective in vigor, and invariably this results in loss of vigor in the offspring. Breeding from pullets or immature stock is another source of lack of vigor in the offspring. An immature fowl cannot impart great vigor to its offspring and the continued breeding, generation after generation, from pullets will result in smaller, weaker and shorter lived fowls. I have experimented along these lines myself, and I find that two or more years of age and still vigorous is the best age for reproducing vigor.

Forcing for egg production by heavy feeding during fall and winter will also impair vigor.

The most vigorous breeding stock is necessary to maintain the vigor of the flock, for "like begets like." The breeding stock should be selected for months ahead of the season, housed and fed for vigor instead of being forced for heavy egg production for the market.

Keeping the young in limited space, in large numbers, is a cause of lack of vigor. The young flock should be culled frequently, the sexes separated as soon as distinguishable, and the pullets sorted according to size, keeping only for future breeders those that show early and rapid growth and development.

The lack of exercise is a cause of low vitality, and slow growth in the crowded pens. On the farm, where the chicks have liberty and a good range, they have also exercise, but in our small yards it should be provided by giving them a good scratching pen, kept well supplied with clean straw or hay, and in this the grain should be buried. Too much to eat and too little to do results in a lack of vigor in the growing stock, and, worse still, brings infertility in the eggs of the breeding stock.

In mentioning eggs, I believe that carelessness in the handling and care of eggs for incubation is not only a great factor in the lack of vigor in the chicks, but also is a cause of the poor hatches of which we hear so much. Our New Zealand cousins have made a series of experiments in this line, and after testing the eggs of thirty-nine different breeds, find that eggs keep best for hatching in a temperature of from 50 to 60 degrees, that below 40 degrees or above 70 degrees the germ becomes weak—dies in a short time. To insure sturdy chicks eggs should be fresh, from a week to not over two weeks old.

Brooding and rearing chicks in insanitary, crowded conditions results in low vitality, and though it is important for the parent stock to be kept in healthy condition, it is equally important that the chicks upon whom our future hopes depend be raised naturally on the best of rations and on free range.

The invariable rule for attaining vigor in the flock should be to eliminate all weak stock whenever we see it. This holds good at all ages, from the baby chick to the mature fowl. The chick which shows weakness at any time should either be killed immediately or be distinctly marked and kept apart from its strong brothers and eventually if it recovers be marketed. The chick may recover from its weakness but still retain the inherited tendency and transmit it to the offspring, so that the rule should be to always eliminate the weak. A satisfactory way of marking that I have used is to paint a bar across the back with a few drops of bluing or with Diamond dye, which will last until the chicken moults.

"How can I distinguish the weak from the vigorous fowl?" asks a correspondent.

The weak chicken is inactive, and dumpy, is inclined to squat down instead or stand, or has leg weakness—does not scratch—is the last to get off the roost in the morning and the first to go on at night. It may frequently be found on the perch during the day, disinclined to do anything, is "born tired."

A loud and hearty crow is one sign of a vigorous male, also his calling up his harem to eat the best and choicest morsels before partaking himself (the careful attendant should see that the male bird that is so devoted to his wives should have extra food himself, or he may fail to transmit his vigor to his offspring).

The shape of the body of the fowl at all stages of development is an indication. A vigorous fowl will be sturdy of frame, with a thick, compact body, large fluff, smooth, bright feathers, prominent eyes. Whilst a fowl that is lacking in vigor will have a long, thin, flat beak and head, a thin neck, slender body, thin thighs and shanks, long, thin and sometimes crooked toes, also usually a tired look.

THE ONE-DAY-OLD CHICK TRADE

The one-day-old chick trade has come to stay. This may be said to be a separate and rather new branch of the chicken business, but it has passed its experimental stage, and both in this country and in England it is becoming popular. It can scarcely be said to be a new business, because it has been known and practiced in Egypt for thousands of years, in fact, it is the only way known there of raising chickens. As soon as one of the large hatcheries there hatches out the chickens, notice is sent to the surrounding villages, and the twenty or forty thousand little chicks are sold within twenty-four hours, or before being fed.

The one-day-old chick trade is, as its title indicates, the selling of baby chicks the day they are hatched. There has been and still is wide discussion over this business, which at first met with but little encouragement from the breeders of fancy poultry, some fanciers averring that it will injure the sale of their fancy eggs, while others even threaten to call in the humane society to prevent such cruelty as selling chickens at so tender an age.

Some of our long-headed fanciers, both men and women, finding there was a demand for one-day-old chicks, rose to the emergency, doubled the price of their eggs in live chicks, and have made a great success of the business. I have had letters from Nevada, Montana, Arizona, New Mexico and even from Old Mexico and Texas, telling of the great success poultry raisers have had in those distant places, raising the chicks after their long journey from Los Angeles, one man writing that he had raised 88% and another 90% to maturity.

L. Yarian of Lima, Ohio, writes: "No branch of the poultry business is attracting more attention at present and no branch of the poultry business is more worthy than the selling of day-old chicks, with hundreds of others in all parts of the United States. I believe it is the best branch of the poultry business ever originated."

Day-old chicks or chicks taken direct from the incubator and securely packed, can be safely shipped to all parts of the United States, except to a very few places, located in some out of the way place where the chicks would have to travel for more than three days.

Occasionally a chick may die en route, but don't they die for you at home, when they are only a couple days old? Certainly they do, and what proof can be advanced that the same chick that dies en route would not have died at home? Is it a cruel practice? I answer emphatically, No. Then some people will ask, what will the chick eat while on the trip? I reply, nothing, because the last thing the chick does before it leaves the shell is to absorb the yolk of the egg, which is Nature's own food intended to furnish nourishment for the baby chick until its little digestive system gets in good working order and is able to handle the food properly.

Poultry men of experience are all agreed that more little chicks are killed by too early feeding than by delay in feeding, and all advise that the chick be not fed until it is at least two or three days old. In fact, some people attribute the diarrhoea of little chicks to too early feeding. If you overcrowd the chick's digestive system before it is ready to digest, you will have bowel trouble, and you know with that you will not have the chicks very long. If it is the advice of men of experience, not to feed until at least the chick is a couple of days old, then why cannot the bird be traveling during that time, comfortably packed in a warm box? That chicks can be safely shipped, has been successfully proved through all who have ever attempted to do so, unless the chicks have very low vitality. Thousands are being shipped all over California and the neighboring states, most successfully, where if eggs had been expressed instead of chicks, many would have been broken en route, for they would have been handled many times rougher than the baby chicks. It would be a very hard-hearted expressman who would throw a box of baby chicks across an express car as they sometimes do when they handle eggs. The selling of day-old chicks should be encouraged, especially among amateurs who often get so discouraged by having poor hatches that they give up after their first attempt.

I have frequently had persons write to thank me for sending the chicks, saying that the chicks arrived in such good condition after three days' journey that they were better and stronger than those hatched at the same time that had not taken the journey. One man in particular, in Mexico, ordered fifty chicks and his success was so great that the neighbors around ended by getting two thousand last season, and this year others in the same neighborhood are already sending for them by the thousand. The day-old chick business has come to stay in America as well as in Egypt.

I want to emphasize the necessity of caring for the chicks immediately on arrival. More chicks are lost or injured in the last stages of their journey than in all the rest of their trip put together.

Find out what train they are coming on, and meet them if possible, or if a telephone is available, have the agent call up on arrival. Make friends with the agent, and tell him that you are getting chicks in and ask him not to pile the boxes, but place them in the shade, but out of the draft.

A good way to make friends with an agent is to go to him for a money order when the chicks are bought, and this gives you a fine chance to talk to him.

SUMMER WORK

Summer is our time for rest from hatching and now our energies must be directed to safely carrying through the summer the brooder chicks and helping the older hens to shed their old clothes and come out in fine and glossy raiment as expeditiously as possible.

Let us first look over our youngsters and see how we can keep them growing. They need a motherly and watchful eye and ear, and a watchful nose also, as much as children do.

Our own lives are made up of little things, but a little chick's life is made up of infinitely little things and it is through little things that success is attained or failure courted. "Be sure to keep the pullets growing," was the vague order given in one of the poultry books that years ago I was studying. The author did not tell

TEN WEEKS OLD PULLETS AT ROOST IN DAVISON OPEN-FRONT COLONY HOUSE.

how to keep them growing nor did he mention what would prevent them growing, and I just hated that man, but since then I decided that, poor fellow, he most likely did not know himself and was only dealing in generalities to write a plausible article for his book or paper without definitely saying anything. But he was right; we must keep the chickens growing, and at the first indication that their growth has stopped we must investigate and find out the cause.

What are the chief causes of chickens not doing well in the summer? Lice and mites. If your chickens are not doing well, treat them for lice, even if you cannot see them, and give their house a good spraying with kerosene emulsion and a little carbolic acid.

Comfort and proper food are the two great factors that will promote the growth of our chicks, and cleanliness is the first requirement. The drinking vessels at this season of the year require special care; whatever may be used should be kept scrupulously clean.

I find a sink brush is an excellent thing for scrubbing out the drinking vessels. They must be kept in the shade. They can be placed in a box set on its side or under a shed or tree, and besides being shaded, they should be frequently replenished during the day.

Sunshine and Shade

Provide shade for the growing chicks; shade from the burning rays of the sun. Nothing is more conducive to health than sunshine, but it must be tempered by shade. Trees and bushes supply the best shade, as the temperature close under growing green leaves is several degrees cooler than under anything that is dry or dead. Few realize what a necessity shade is to fowls.

If an epidemic seizes the half-grown chicks, it is attributed to any cause on earth but the lack of shade, when, in very many cases, this is the sole cause. Vertigo, blindness, stunted growth may all be due to the glare of the sun on unsheltered yards. Shade is a necessity, and if trees or shrubs are lacking, a good shelter can be made by driving a few stakes or small posts into the ground and making a frame upon which palm branches or brush can be laid. I have found a very serviceable temporary shade can be made by ripping open a common gunny sack and nailing four laths on the edges. This little frame can be laid across the top of a small pen or even hung on wire fence and afford a grateful shade.

Overcrowding or the chicks huddling for even one night may stunt the growth or be the means of bringing on an epidemic of colds which may result in roup.

But how to stop them crowding? A mother hen often solves the difficulty by taking the half-grown chicks on the perch with her, but for brooder chicks some other plan must be found; the best way is to divide them into flocks or colonies of only twenty-five in each, and supply comfortable perches for them. The chicks will in a short time take to the perches of their own accord.

At one time I had not enough colony coops and a great many chicks. I put them a hundred together in my regular henneries, but they crowded and I not only was losing every night some of the best, but the survivors looked very badly. They sweat off in the night all they had gained during the day. I realized that this meant failure for me if I could not control it. I spent my evenings going around and patiently placing the chicks, hundreds of them, on the perches till I was completely tired out, when I decided to make it so desperately uncomfortable for them they could not crowd.

I bought a bundle of six-foot lath and made a lath platform or floor, by nailing them one and a half inches apart, the width of a lath, on stringers one inch by three. This made a flooring of small lath perches three inches above the ground, and made it so uncomfortable for the chicks to crowd that it entirely prevented it. I placed regular perches four or five inches above the lath floor and in a few nights, on making my nightly rounds with my lantern, I had the satisfaction of finding all the chicks on the regulation perches. I have recommended the lath platform or floor to many and it has proved always successful.

Teaching Them to Roost

It is sometimes difficult to persuade the young chickens at this time of the year (September), when moved to winter quarters, to go into the coop or house, which they should occupy. The little perversities insist on returning to the place where their mother has raised them, or they will huddle together on the ground, while the older ones fly into the low trees. Night after night, they have to be carried to their house. I, however, have found that by driving them gently with a broom for two or at most three nights, they will soon learn what is expected of them. A broom is by far the best way of driving chickens without frightening them.

A broom in each hand is the best way of driving a large herd of turkeys, also, by gently waving them on each side. They will be afraid of the broom, but never become wild or afraid of the attendant in this way. It is entirely possible to drive the profits out of a flock of hens by stoning and pelting them every time they get into mischief. Be quiet in your manner if you wish to be successful with hens. Make the fowls feel that, when you are present there is a protector among them, not something that is likely to scare or harm them. The only way to keep your fowls on good terms with you is by keeping them tame and treating them in a common-sense manner.

Protecting Chicks from Older Fowls

It may sometimes be necessary to allow young and old fowls to run together. This creates trouble, as the young chicks require more frequent feeding than the older ones. To avoid this trouble, make a pen about six feet square and covered with wire netting. The pen should be made on a framework so that it can be easily moved. Feed for the chicks is scattered in the pen far enough from the edge so the older fowls cannot reach it from the outside. Then the pen is raised on blocks, just high enough to allow the chicks to pass under but will prevent the older fowls from getting inside.

The Dry Hopper

In the matter of feeding hens on a farm, I would much prefer the dry hopper method, keeping one hopper full of mixed grains and one hopper with beef scraps or granulated milk, and letting the fowls have free range until it is time to put them in their winter quarters. Then, instead of only grain in the hopper, make the mixture of bran, corn meal and alfalfa meal, or take one of the good balanced rations sold at the poultry supply houses for the hopper. The reason for this change which should be made gradually, is that the fowls being confined, do not get the exercise and consequently may get overfat from eating the whole grains, while the finely-ground food has to be eaten more slowly. For fowls in confinement, besides the hopper or finely ground feed, they should have a scratch pen in which the grain is thrown every morning for them to scratch in. This will give them the exercise which they would otherwise miss after being on free range all the summer.

After getting the fowls accustomed to their winter quarters, you can, if you wish, let them out for two hours before sundown to run on the grass or green winter wheat or alfalfa. This will give them a little exercise and change, but it is not absolutely necessary unless quite convenient. Of course, they must be supplied with green food and a balanced egg ration.

BROILER RANCHES

Broiler raising is one of the lucrative branches of the poultry industry. It is a business, however, which should not be entered into without study or experience. There are some very large broiler ranches in the neighborhood of Los Angeles.

The ration for broilers is usually that given for chicks till they are four or five weeks of age, when they are finished off with a fattening ration for from two to three weeks. The average cost of raising a broiler is from fifteen to eighteen cents, while the selling price on contract is from fifty to sixty cents at a pound and a half in weight.

By using the ration given for broilers after the first two weeks, some breeders have attained the weight of two pounds for their broilers at six weeks of age. This was in small lots of twenty-five to fifty broilers in a brooder.*

* See page 39.

THE TRAP-NEST

Trap-nests are one of the inventions of this progressive age. It is the surest, quickest method of securing better eggs and more of them. A trap-nest is a nest box, the entrance to which closes automatically when the hen steps into the nest and keeps her in the box until the person in charge releases her, thus showing which hen laid the egg.

The progressive farmer or dairyman knows that he must test the milk of his cows and he finds when he begins to do so that he has cows in his herd that do not pay for their keep. It is the same in the poultry business; in every flock of hens there are idlers that do not pay for their feed—they lay so few eggs that their owners are out of pocket by keeping them. I would not have believed this had I not discovered it to be the case with some of my own hens. The first season that I used trap-nests I found a hen which went on the nest every day, but only laid four eggs in one month, while another in the same yard laid twenty-nine. It was a revelation to me. The first year I discovered that nearly one-fourth of my hens barely paid for their board. That was not the kind of hens I wanted. I was in the business for profit and not loss, so I weeded them out, and very good eating they made.

The second year I got, with a reduced flock, a twenty per cent less feed bill and fully twenty-five per cent increase of eggs—more eggs at less cost. Surely the trap-nests repaid me for the slight extra trouble of attending to them. They were not only of use in discovering the best layers, but I became better acquainted personally with each hen. I found that the hen which laid the most eggs had the most fertile eggs, while the poor layers' eggs were not nearly so fertile.

Trap-nests make the hens tame and tame hens lay more eggs than wild hens. Some hens may at first object to being handled, but after a few days they become reconciled to it. My White Plymouth Rocks were so tame that when I opened the door they would step into my hands or sit quietly until I lifted them up to ascertain the numbers of their leg-bands.

In order to make the use of the trap-nests efficient, we must be able to know each hen individually, and for this purpose each hen must wear a leg-band, a small bracelet, made of copper or aluminum with a number on it.

By means of the trap-nest one can discover any hen that is becoming too fat, or too thin, and she can be moved into another and more suitable pen. The trap-nest also renders a great service in detecting the egg eater. If there is reason to suspect a certain hen of this villainous habit, give her an egg while she is on the nest; if the egg after a time disappears it is pretty good evidence that the culprit has been discovered, and decapitation should be the verdict.

Another advantage in using trap-nests is that it gives one an opportunity to examine the hens for vermin, and by taking a small can of insect powder around occasionally while visiting the nests.

and powdering the hens, they can be kept perfectly clean with very little trouble. I use a baking powder can, having perforated the lid, making a large pepper pot. A liberal use, not blown on out of an air gun, but freely peppered on the hens, is very beneficial.

I visit the nests about three times during the morning to release the hens and gather the eggs. One trap-nest is required for every three hens. When a hen is taken from her nest, the egg is marked with her leg-band number and the date and credit is given her on the record sheet or record book. This is a sheet or page marked off in squares of thirty-one days with the hen's name or number at the head of the line. I mark B for broody, S for sold, M for marketed, and so on, and have in this way the history of each hen at a glance.

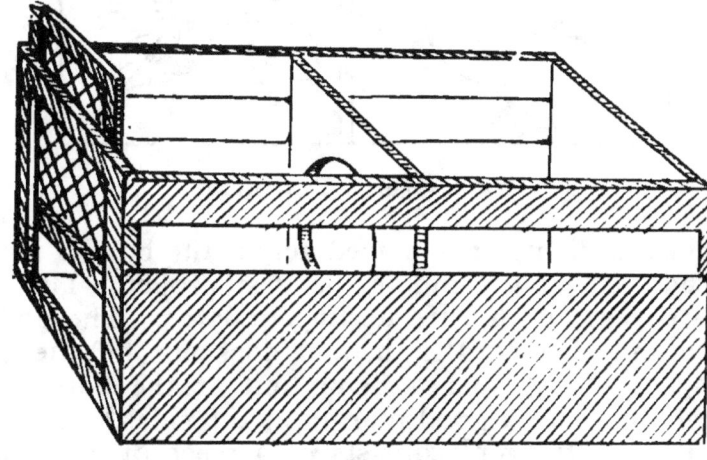

MAINE EXP. STA. TRAP-NEST, IMPROVED.

Trap-nests have taught me which hens lay the best shaped eggs, which the largest size, which the strongest fertilized, which are the best winter layers, which pullets begin early, the number of eggs they lay in succession, the number of times they become broody and many other facts that can be learned in no other way; in fact, I find my records exceedingly interesting and profitable reading. Trap-nests were a perfect revelation to me and aided me in my success with poultry.

There are a number of trap-nest plans, also trap-nests, on the market, ranging in price from $1 to $25. I have bought and tried several, and find that the most satisfactory trap-nest is one that has two compartments, one the nest where the hen lays the egg and the next where she waits to be released. I find that if there is only one compartment the hen often stands upon the egg and is apt to break it accidentally when she wants to come off and so learns to eat eggs. The two-compartment nest is comfortable for the hen and convenient for the attendant.

The nest box here described is adapted from the first that was made at the Maine Experiment Station, is very simple and easily made. I use wooden cracker boxes or shoe boxes, and easily made two in a morning. The wire was a little difficult to bend, but a boy bent it for me.

The nest box is very simple, inexpensive, easy to attend and certain in its action. It is twenty-eight inches long, 14 inches wide and 14 inches deep. A division board, with a circular open-

·HEN ENTERING TRAP-NEST.

ing 7½ inches in diameter is placed across the box 13 inches from the back end. The back end is the nest proper. The door is a light frame covered with wire netting or laths. The door is 10½ inches wide and 10 inches high, and does not fill the entire entrance, a space of 2½ inches being left at the bottom and 1½ inches at the top, with a good margin at the sides.

The "trip" or "trigger" consists of a piece of wire about 3-16 of an inch in diameter and 26½ inches long, bent as shown in drawing. To hold the trip wire in position and let it roll sidewise easily, nail a cleat to the cover and put two staples for the wire to hang in near its bent corners. The long end of the wire hangs in front of and close to the center of the 7½-inch circular opening. When the door is set, the half-inch section of wire marked "A" comes

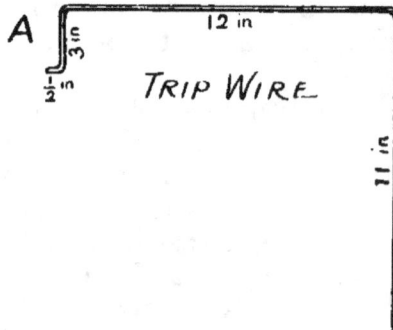

TRIP WIRE.

under a hardwood peg or a tack with a round head which is driven into the lower edge of the door frame. The hen passes into the nest through the circular opening and in doing so presses the wire to one side and the trip slips from its connection with the door.

The door promptly drops down, without noise and without frightening the hen. The double box, with the nest to the rear end, is necessary, as when a bird has laid and desires to leave the nest she steps to the front and remains there till released. With one section only she might crush the egg by standing upon it.

One word of caution: It is well to have nests enough, because the hens must be coaxed to lay, and when they are ready they must not be kept waiting. If a hen is dissatisfied with her nest she may hold her egg for twenty-four hours, and in time be taught to lay only every two or three days. It is wise to encourage the hens to lay, and I have found these trap-nests are much liked by the hens whilst others that I bought frightened them and prevented them laying, entirely defeating the object of the nest, which is more eggs and better hens.

There is not any patent on this nest; any one is at liberty to make and use it, as it is the original trap-nest made by the Maine Experimental Station, slightly altered.

GRIT AND GIZZARD

One of the most important things necessary for the health of poultry is a supply of grit of the right kind. Nature provides a use for every organ of the body, and in every body an organ for each specific duty. Most animals are provided with teeth to enable them to prepare their food for the action of the fluids secreted by the stomach, pancreas and liver. It will also be remembered that besides being crushed in the mouth by the teeth, the food is acted on by the saliva.

Nature has not endowed birds with teeth, but it has provided a good substitute in the gizzard. This is a tough, strong, muscular organ, so situated in the body that all food taken into the mouth must pass through it. Previous to passing through the gizzard, all food has been received into a pouch or bag, the crop, where it remains some time. There it is soaked with and acted upon by a fluid secreted in and by this pouch, and a modified process takes place similar to that of the saliva in the mouth of animals with teeth.

The food gradually leaves this pouch (the crop), passes through the proventriculus and into the gizzard, where it is ground up, and thence it goes to the intestines, where, after being mixed with other fluids, it passes on and the nutriment is absorbed. No doubt a bird may be made to exist for a time, perhaps a considerable time, without grit, just as a person may live for years with bad teeth, or perhaps with none at all. We all know how little such people enjoy their food or health, and surely if the birds do not have the means of masticating their food they can neither be healthy nor enjoy their food, and will not give their owners a good return for their food and care.

The Best Grit

The gizzard is a marvelously strong little mill and when provided with the proper grit, or little grindstones, will keep the fowls in good condition. Hard, sharp substances are necessary, such as flint stones or granite pounded up. Broken china, earthenware, glass and all such substances broken up make excellent grit.

When the grit has not sharp edges, the harder parts of the food are not digested, husks and green food accumulate and frequently 'cause a stoppage between the crop and the gizzard, so that nothing but liquid can pass. A lack of sharp grit brings on diarrhoea; also, the gall overflows and sometimes the gall-sack bursts. There are two passages, one into, and the other out of the gizzard; they are both on one side of it. The one leading out of it is much smaller than the one leading into it. Thus the gizzard can receive larger substances but cannot get rid of them until they are ground small; and sharp grit is needed for this.

When I first came to California I purchased a grist mill and, alas, I had broken china also! I had two dozen hens just bought and proceeded to grind up some crockery for them. The man who was building my fence thought it dreadfully cruel of me, remarking

"It's enough to kill a dog; let alone those poor hens." "The hens will not eat it unless they need it," was my reply, though I agreed with him about the dog. To his surprise those hens ate almost a quart of it. None of them died and they soon commenced to lay. Give the little chicks the small chick-grit. Eight pounds of this will be sufficient for the first two months of the life of fifty little chicks and then they should have a larger size. One hundred pounds of hen grit, which can be bought at the poultry supply houses, is sufficient to last a hundred hens about a year.

Pigeons consume more grit than hens, proportionately to size. Give pigeons grit to keep them healthy. My attention to grit and gizzards was aroused many years ago. "Will madame look to what I have found in the interior of this fowl?" said my French maid to me. She had opened the gizzard of a fat young hen and had found thirteen china buttons and two pearl buttons or parts of them, mixed with the black adobe mud. Since that day I have tried to keep my fowls well supplied with grit.

Starve for Lack of Grit

"I cannot think what ails my fowls," said one lady. "They have all the food they can eat, but here is another dead." "Have you ever opened one to discover the trouble?" I asked. "Yes, but I never find anything." "Well, I think your fowls have indigestion," I said, "but we will hold a post mortem on this one and try to solve the difficulty." We found a medium sized gizzard, full of dark earth, no stones, no grit, not even buttons. That told the story, the fowls were starving to death in the midst of plenty, just for lack of grit to grind their food.

I occasionally make curious discoveries when I hold a post mortem, for the contents of a school boy's pockets are scarcely more varied than those of a fowl's gizzard, when not supplied with the proper kind of grit. My Indian Runner ducks, being great pets and never doing any mischief, were allowed the freedom of my place. I had noticed them around the out-door fireplace where the cauldron was boiled, old boxes, building scrap and rubbish being used for the fire.

I thought the ducks were picking up bits of charcoal, but one morning I found a fine duck dead. The post mortem revealed an enormous gizzard, twice the usual size, on opening which I found a number of nails, some bits of wire, two two-pointed tacks. Several of the nails were embedded in the gizzard and the largest one pierced quite through it. The ducks had always been supplied with plenty of river sand, but this particular duck seemed to have developed an ostrich's appetite. After that I gave them also the smaller chick grit and with most excellent results, for never ducks laid as many eggs as did those. Grit, oyster shells, or clam shells, and charcoal are indispensable for fowls.

The Symptoms of Grit Craving

When your hens seem "mopey" just break up some old china, and see if they will not refuse the best food for it.

When you see water run from a hen's mouth, when she puts her head down, the trouble is indigestion. Give her grit and charcoal.

When your hens do not care for their food, tone up their appetities by a dose of grit.

When they are not laying as well as you think they should, give them grit.

When hens moult slowly, it is often from impaired digestion. Give them grit and charcoal.

When you want the hens to derive all the benefit of the nutrition in the food, supply them with good, sharp grit.

If you want vigorous, profitable hens, give them a liberal supply of grit.

When your hens are too fat, when they lay thin-shelled eggs, give them grit.

A friend of mine was very much troubled with soft-shelled eggs. She got her husband to take his wagon to the hills, where there is a good quarry of what is called rotten granite. He brought home a load of it, and in a few days the hens laid hard-shelled eggs and she told me that the shells were so hard that the chicks could hardly break out of them.

The value of good sharp grit can scarcely be overestimated, and yet even intelligent people do not realize it. Some think that there is grit enough in the natural soil. This is rarely the case, for hens, wild birds, or pigeons pick up the sharpest and best grit, so that even on a farm where the hens have free range, there is rarely enough grit of the proper kind, and when fowls are kept yarded there is never enough unless they are artificially supplied. If you doubt this, try the experiment of giving your hens some broken china. The pieces should not be larger than a pea and should have three sharp corners. You will be surprised to see how eagerly the hens will eat the china.

The best layer I ever had laid 225 eggs in nine months and moulted during that time. She was the greatest eater of grit I ever saw. Every night before going to roost she ran down to the grit box and took three pieces. Every time she laid an egg she refreshed herself with some grit, and I learned by observation that all my best layers were the most constant visitors to the grit box. Hens that consume the most grit are those that get the most nutriment out of their food, lay the most eggs, are the heaviest, have the most fertile eggs and pay the best.

Grit to grind the food and charcoal to keep it pure during this process and, for laying hens, oyster shells to supply the lime for the eggshells; these are so necessary that we are almost tired of the mention of them in the poultry papers, but, "lest we forget," I have written about them again.

PESTS OF A POULTRY YARD

Fleas

The common hen flea (pulex avium) is prevalent in the Pacific States. It is found in filthy hen houses, especially those located on sandy soil. Dirty nests, cracks, dust and dark corners are favorite breeding places for them. They produce great irritation of the skin and in young birds the growth may be permanently stunted and many young chickens killed by them.

For treating flea bites, bathe the bites with vinegar and water, or lemon juice, and apply carbolated vaseline or lard in which a little carbolic acid has been mixed—5 drops of carbolic acid (90 per cent) to a tablespoonful of lard.

To free poultry houses and yards of the fleas, use whitewash freely, adding a pint of carbolic acid to every twelve gallons of whitewash. Spray it or slop it thoroughly into all the corners and cracks. Dark dusty places in the poultry yard afford favorable breeding places for fleas. These corners should be soaked with hot soapsuds or boiling salt water to kill the young broods of fleas. Use carbolized lime, tobacco dust and moth balls in the nests.

Bedbugs and Ticks

Bedbugs sometimes attack poultry on their roosts and suck their blood. In California there is also a species of tick that is fatal to poultry which somewhat resembles the bedbug of the East. To destroy them fumigation is usually employed, either fumigating with sulphur, or, better still, the cyanide process used for the scale on citrus trees.

To fumigate with sulphur close every door and window and see that there are no cracks to admit the air. Burn one pound of sulphur for every 100 square feet of floor space in the house. A house 10x10 will require one pound of sulphur; one 20x10, two pounds, and so on. The sulphur must be burned in iron vessels which should be set on gravel or sand, so there may be no danger from fire. Into each vessel put a handful of carpenter shavings saturated with kerosene and upon these sprinkle the sulphur. Apply a match to the shavings and hastily leave the house, closing the door. The house should remain closed for 5 hours. Fumigation may be followed by thoroughly whitewashing the inside of the house. Painting or spraying the house with corrosive sublimate is also very effective. Care must be used in handling this poison.

Mites

There are several varieties of the tiny blood-sucking mites to be found in carelessly kept henneries. The red mite is the most common and active of all parasites which attack birds. It is about one thirty-fifth of an inch in length, white or grey in color, except when filled with blood, when they will be red or black. It hides by day in the corners and crevices of buildings, nests, perches, floors, etc., where they may be found in clusters. At night these clusters scatter over the birds and by pricking the skin can fill themselves with blood. They are injurious not only on account of the blood they draw, but because of the itching pain and loss of rest. They will even kill young fowls and sitting hens. When they are discovered vigorous means should be adopted to get rid of them. The Iowa State Experiment Station gives a full description of the best and cheapest way of exterminating these mites. At this station the kerosene emulsion was found to be perfectly effective in killing them. It is made as follows:

KEROSENE EMULSION—In one gallon of boiling water dissolve one pound bar of soap or one pound of soap powder. Remove from the fire, add immediately one gallon of kerosene, churn or agitate violently for ten minutes, or until the solution becomes like a thick cream. If the oil and water separate on standing, then the soap was not caustic enough. Take one quart of this, add to it ten quarts of water; spray thoroughly the houses every three days with this diluted emulsion until all the mites are exterminated. To make it more effective, you may add one pint of crude carbolic acid to the emulsion as soon as taken from the fire. The diluted emulsion (one part to ten of water) is also used to rid fowls of lice. By using this spray once a month always, the houses can be kept perfectly free from vermin and thoroughly disinfected from disease.

Lice

There are nine varieties of lice affecting poultry. Some of these lice spread rapidly. One infested bird is capable of spreading the vermin through a large flock. They cause dumpishness, drooping wings, indifference to food and may stunt or even kill the chicks. One of the best means of preventing lice is the dust bath. This bath should be a wallow of freshly turned earth, mellow and slightly damp, out of doors under some tree in the summer time, or in a box six or eight inches deep in the hennery in the rainy weather. Provided with a good dust bath, healthy hens will almost keep themselves clean from lice. When fowls are badly infested with lice they should be well dusted with a good lice powder, of which there are a number on the market.

In looking for lice on a fowl, examine the head feathers carefully, one by one, then look under the wings and along the shafts of the underside of the long wing feathers, examine the feathers of the cushion and saddle down to the skin, and then turn the fowl quickly and look beneath and around the vent. If you have eyes to see you will find them. If you find only one or two, a thorough dusting of the bird will be all that is needed, but if the lice are plentiful, a more vigorous treatment will be necessary. Lice breed very rapidly on the fowl among the feathers where the warmth of the bird's body can hatch the eggs, which are deposited singly or in clusters among the soft feathers. They seldom ever breed on young chicks, but are passed along to the chick by some lousy adult bird.

Lice Eggs on Hen's Feather. Should be Pulled Out and Burned.

How to Keep Poultry Free from Lice

The following formula is used at the Maine and Cornell Experiment Stations:

Take three parts of gasoline, one part of crude carbolic acid. Mix these together and add gradually, while stirring, enough plaster of Paris to take up all the moisture, the liquid and the dry plaster should be thoroughly mixed and stirred, so that the liquid will be uniformly distributed through the mass of plaster. When enough plaster has been added, the resulting mixture should be a dry, pinkish brown powder, having a fairly strong carbolic odor and a rather less pronounced gasoline odor.

Do not use more plaster, in mixing, than is necessary to blot up the liquid. This powder is to be worked into the feathers of the bird affected with vermin. The bulk of the application should be in the fluff around the vent and under the wings. Its efficiency can be very easily demonstrated by anyone to his own satisfaction. Take a bird that is covered with lice and apply the powder in the manner described. After a lapse of about a minute, shake the bird, loosening its feathers with the fingers at the same time, over a clean piece of paper. Dead and dying lice will drop on the paper in great numbers. Anyone who will try this experiment will have no further doubt of the wonderful efficiency and value of this powder.

For a Spray or Paint

To be applied to roosting boards, walls and floor of the hen house, the following preparation is used:

Three parts of kerosene and one part crude carbolic acid. This is stirred up when used and may be applied with any of the hand spray pumps or with a brush.

In both of these formulas it is highly important that crude carbolic acid be used, instead of the purified product. Be sure and insist on getting crude carbolic acid. It is a dark brown, dirty looking liquid and its value depends on the fact that it contains tar oil and tar bases in addition to the pure phenol (carbolic acid).

DISEASES OF POULTRY

There is no reason for chickens being unhealthy except, as a general thing, from the carelessness or ignorance of their owners. Carelessness in not keeping the fowls clean, in not being regular in their feeding, in the lack of pure water and shade and in giving them either draughty sleeping quarters or too close and badly ventilated coops.

Poultry keepers in the East, after years of trouble and anxiety over roup, which I really think is much worse there than here, are coming to the conclusion that open-front houses, even there, where they have zero weather, will prevent roup and colds.

Here in our favored climate, open-front houses, cleanliness and plenty of green food are a sure prevention of roup.

I am glad to be able to say that although there are more than double the number of pure-bred fowls in California now than ever before, there is a minimum amount of roup. Poultry raisers are using common sense in the feeding and care of chickens, looking upon poultry raising as a business, a money proposition, when handled in a business-like way, and the result is very little roup and less sickness of any kind.

Roup must be transmitted by contagion; healthy fowls will not have it unless a roupy fowl is introduced into the flock, or the infection is brought in through water or food, through coops in which roupy fowls have been confined or through the infection being carried on the garment of the attendant.

Many Kinds of Roup

It was formerly the custom to call nearly all the ailments of fowls due to taking cold by the name of "Roup." Dr. Salmon of the Bureau of Animal Industry, Washington, D. C., makes a distinction, however, between the different kinds of colds or roup, simple catarrh and infectious catarrh, also called roupy catarrh, and diphtheritic catarrh or diphtheritic roup. Simple catarrh is easily cured, will often get well without treatment; roupy catarrh is very infectious and more difficult to cure; but diphtheritic roup is the worst of all and greatly resembles the diphtheria of children. There is also another disease called "Canker" which much resembles diphtheritic roup, but is less severe. It is caused by another germ and needs other treatment.

Catarrh

All of these diseases commence in the same manner. Usually the first symptoms noticed are a slight discharge from the nostrils, eyes wet and watery from mucus, and often some bubbling at the corners with coughing and sneezing. In simple catarrh more serious symptoms will not have developed in a few days, but with roupy catarrh the discharge thickens and obstructs the breathing by filling the nostrils and there is a foul odor to it. Sometimes swell head develops, then one or both eyes are closed, the birds wipe their eyes on their shoulders, sleep with their heads under their wings and the discharge sticks to and dries on their feathers.

This dried mucus will spread the disease through the flock, for in it are the germs of the disease, the seeds of which may be sown whenever the chicken moves or shakes itself, or when others touch it or a feather falls. Chickens with this disease should be isolated, the mucus gently washed off, using a disinfectant in the water, a few drops of carbolic acid or a tablet of protiodide of mercury in a pint of water. Roupy catarrh is difficult of cure, is very infectious and often fatal.

Diphtheritic Roup

Diphtheritic roup is the worst of all. It requires different remedies to the simple catarrh or roupy catarrh. It commences usually in the same manner with a slight cold, but the mucous membrane of the mouth, throat, nasal passages, and the eyes are affected. False membrane forms on these parts, very much resembling in appearance the diphtheria of children, and by some thought to be the same. At first the patches are small and scattered but have a tendency to run together. The disease appears suddenly, the fowl is feverish, dumpish and disinclined to eat. As the disease progresses the mouth and throat become filled with false membrane and mucus until the fowl dies of suffocation, or the poison from the disease gets into the circulation and the fowl dies of blood poisoning or paralysis.

Canker

Canker is sometimes confounded with diphtheria. It is an ulcerative disease of the mouth. It is frequently found in cock birds after fighting and is common in birds that have been working in mouldy or musty litter or that have been fed on spoiled grain. The disease is seldom noticed until the fowl shows a collection of yellowish ulcers or cheesy growth on the roof of the mouth, the side of the tongue or the angles of the jaws, and sometimes at the opening of the windpipe. It is very common among pigeons.

Roup cures can be bought at the principal poultry supply houses, but for the use of those living in the country too far away to procure these, I will give a few simple remedies that can be easily and quickly used in the first stages, thus arresting an epidemic. For local treatment a good atomizer is the most satisfactory way of applying it, or a small syringe, and as handy as anything is a small sewing machine oil can.

Remedies

(1) When first the cold is noticed, put a bit of Bluestone (sulphate of copper) in the drinking water. A piece as big as a navy bean in a quart of water, not any stronger. This is a germ killer, dries up the cold in the head, is a disinfectant and will prevent the other chickens taking the disease. So if any chick takes cold, put this into the water of the whole flock for a week to prevent the disease spreading.

(2) For a Common Cold: A pill of quinine and one of asafoetida (1 gr. of each), with half a teaspoonful of cayenne pepper

will frequently cure a cold in one night. Aconite also is a good remedy. One drop in a teaspoonful of milk. Always give a grown hen the same dose as to adult human being.

The following are cures for Roupy Catarrh:

(3)   One tablespoonful of castor oil, half a tablespoonful turpentine, a tablespoonful of kerosene, a tablespoonful of camphorated oil and four drops of carbolic acid. Shake before using. Squirt a drop up each nostril and into the cleft of the mouth, and for swell head rub the whole head with it. This is an excellent cure and cheap.

(4)   Put one cupful of kerosene in half a gallon of water; the oil will float on top; dip the fowl's head slowly into this, holding it under whilst you count three. It will sneeze and cough and you must wipe off all the mucus with a rag and carefully burn the rag. Repeat the treatment twice a day.

(5)   Take of lard two tablespoonsful; vinegar, mustard, cayenne pepper, each one tablespoonful; mix thoroughly, add flour enough to make a stiff dough. Give a bolus of this the size of the first joint of the little finger. One dose frequently cures. If not, repeat in twelve or twenty-four hours.

(6)   Dr. N. W. Sanborn gives as a remedy: "Spray all mucous surfaces with the following: Extract of Witch Hazel, four tablespoonsful; liquid carbolic acid, four drops; water, two tablespoonsful. Do this twice a day, squeezing the bulb of the atomizer five times for each nostril and twice for the mouth. If there are any watery or foamy eyes, give one squeeze for each.

(7)   One part of pulverized gum camphor and seven parts of pulverized liquorice root. Blow up the nostrils, into the cleft of the mouth and down the throat. This should be made fresh, as the camphor evaporates.

(8)   Five drops of eucalyptus oil on a bit of bread or a lump of sugar night and morning.

(9)   For Diphtheritic Roup: Peroxide of hydrogen is, I think, the best remedy. Dilute with from one to three parts of water. The solutions, when applied to diseased surfaces, begin to foam, and should be repeated until there is no more bubbling. A little of the solution forced into the nostrils by the use of a dropping tube or atomizer is driven higher up into the nostrils by the force of the foaming, reaching parts otherwise out of touch.

(10)   For Canker: Four grains of Sulpho-carbolate of zinc to one ounce of distilled water. Paint the canker spots with this night and morning and in three days the germs will be destroyed. The chickens should have nourishing food, such as bread and milk and chopped onions.

If you have any doubts as to whether the disease is canker or roup, you had better use the peroxide of hydrogen one day and the zinc the day following, alternating the treatment. It will not do to mix the two medicines at the same time, as one neutralizes the other.

Kileroup, a patent medicine which is very highly extolled and can be bought at any supply house (see advertisement at the end of the book).

The Diseases of the Lungs Are Bronchitis

Bronchitis: Bronchitis is confined to the lining membrane of the breathing tubes. Bronchitis is caused by exposure to storms, especially when birds are housed in too close or too warm a building, or by sudden atmospheric changes, direct currents of cold air, irritating particles of dust or lime, or by spreading of inflammation from catarrh of throat or nostrils. It is not considered infectious, though it may be almost epidemic from the same cause affecting several of the flock. Birds sent on trains to an exhibition or to a new owner sometimes develop bronchitis. The hot, close air of the show room and the warm corner of the express car, succeeded by exposure to wind or cold, very frequently develops bronchitis.

Symptoms.—There is from the first a rise of temperature, and a little difficulty in breathing. The lining membrane of the bronchial tubes is dry and swollen, hindering the passing in and out of the air. On listening to the respiration, a whistling sound may be heard; later on a rattling or bubbling sound, caused by the air passing through accumulations of mucus, is heard.

Treatment.—Place the affected birds in a comfortable and reasonably dry place, where the temperature will be even. Give soft and cooling, but nourishing, food, such as bread and milk. Give one drop of tincture of aconite every three hours, and two or three times a day a half teaspoonful of honey with five drops of eucalyptus oil. Kileroup is also a good cure for bronchitis.

Roup, Cure For.—The following treatment for roup, when it has extensively infected the flock, is recommended by the New York Experiment Station: A solution is made of one teaspoonful of permanganate of potash, dissolved in one pint of water. All the cheesy matter is picked off with a toothpick and the spots painted with iodine. Then the heads of the sick fowls are dipped in the solution. This treatment to be repeated daily until a cure is effected.

TOWN LOT FOWLS

The rear of a city lot can be made to yield both profit and pleasure when devoted to poultry and fruit trees, and many families may enjoy fresh eggs and an occasional roast chicken, or a "Christmassy" chicken pie by simply utilizing some of the vacant space in the rear yards of their homes.

We sometimes hear that chickens cannot be raised successfully on a city lot because the land is too valuable and that the business will not pay where all the food has to be bought.

The value of a city lot is often overestimated when chicken raising is suggested for the back yard, but the question is, what income is your back lot now yielding?

I expect that the majority of city back lots are either an outlay or an eyesore to their owners. They grow nothing but grass or weeds, for which nothing is received. When mowed there is that expense to it, with the water tax added, which is not inconsiderable.

As much as I like lawn and flowers in the front of the house, I think the ofttimes neglected back yard should be made valuable also. Nothing to my taste can improve it like fruit trees, which are benefited by having poultry around them, and will bring in good returns, as I know by experience.

The main requisite to making a success of poultry raising on a city lot, or anywhere else in fact, is to be thoroughly in love with your fowls and your trees. The man or woman who hates to work around the hens, who grudges the time and trouble, will never make a success of the work and had better let it alone.

How to plan your back lot? It should be fenced to suit your space and poultry. If it is a small yard, it may be difficult to fence it high enough for the active breeds, such as the Leghorns, but if you use poultry netting and do not place any rail on the top, you will not have any trouble with the American breeds, even with a comparatively low fence. If there is no rail on the top, the fowls do not see where the netting ends and they seldom try to find the top, but with a rail they light on that and over they go.

It may help a beginner to see the plan of my chicken yard on a city lot. The chicken yards are 50 feet by 32 feet; there are eight fruit trees and three water faucets in the yard. The fruit trees, plum, peach and fig, yielded several dollars' worth of fruit two years after planting, and as they grew older, increased the value of the crop in the back lot, and gave the fowls shade.

Hen House Construction

The earth around the trees is kept well spaded and moist, so the hens enjoy it as a dust bath and that keeps them clean from lice and mites. The hen house is a shed thirty-two feet long and eight feet wide. It is divided in two parts for two pens of fowls. Each end of it is composed of a roosting room eight feet by eight feet, with space enough for forty hens, if necessary, although I never wish to keep more than twenty-five in each side.

The roosting room is separated from the scratching pen only by a board twelve inches wide, to keep out the straw. The back and sides of the roosting room are of tongued and grooved flooring and perfectly tight. The whole length of the front of the shed is open, except the roosting room, which has a front of burlap. One side of the roosting room is entirely open into the scratching pen, so that the roosting room is only tightly enclosed on two sides and has free ventilation into the scratching pen and only the burlap on the south side. Consequently my fowls never have colds. The roof is of shakes twelve inches to the weather. The back of the shed is six and a half feet high, the front five feet.

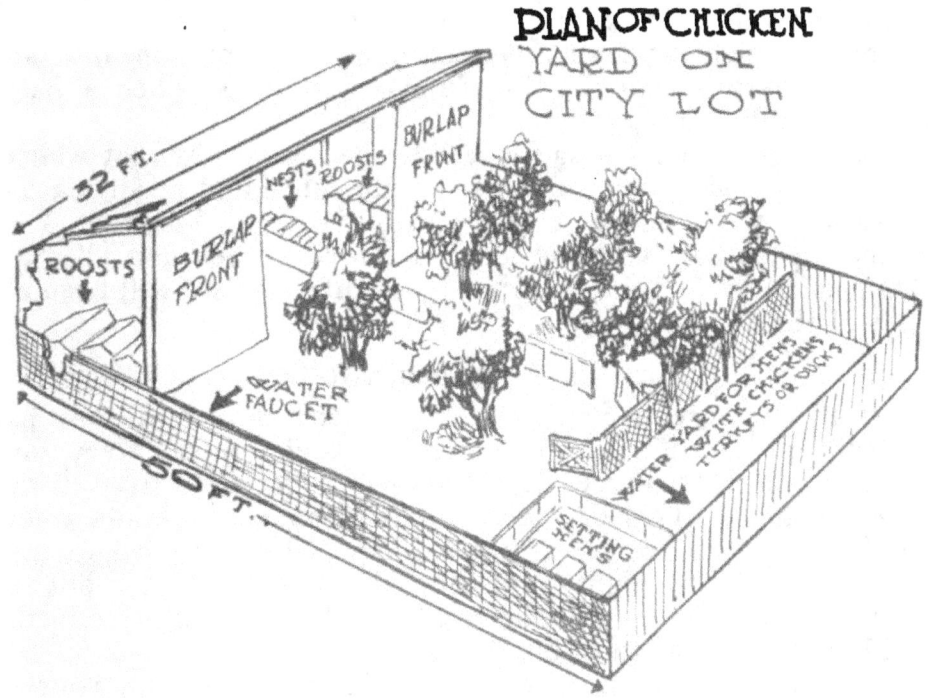

At the south end of the two yards is a smaller one for sitting hens or for young chicks, as they do better kept away from the older fowls. This small yard is very useful for fattening chickens, turkeys or ducks for the table, and in it I have a small portable coop for the youngsters.

I have a water faucet in each yard. This is a great saving of labor and anxiety, for if I am to be absent any length of time I leave the faucet dripping just a little and know the hens will not go thirsty.

I feed grain in the scratching pens, dry mash in hoppers, green lawn clippings and refuse vegetables, besides the table scraps.

There is a saying that an American family wastes or throws away food enough to support a French family. Why not give all this waste to some hens? The table scraps, the scrapings of the plates, the outer leaves of cabbages, even the parings of potatoes, apples and nearly all vegetables now consigned to the garbage pail would be enough to almost keep a few hens.

Possibilities of a Town Lot

Have you any idea what returns one dozen laying pullets or hens would give you? I have, for I have kept that number on a town lot. I have not an accurate account of all the eggs laid, but I know there were over two thousand in one year, more than enough to supply a family of six with delicious fresh eggs and to raise between fifty and sixty young fowls for frying and roasting, besides the old ones for stews for "poulet au ris," a French dish of which we are extremely fond.

Nine-tenths of the home owners have sufficient space in their back yards to produce enough chickens and eggs to supply their own families, and in this way greatly lessen the expense of living, or in other words, make enough to pay their meat and grocery bills, or else give them all the fresh eggs they can consume with a nice fry always available for Sunday dinner or when a friend unexpectedly drops in.

I will give you a formula for feeding hens on a town lot which I will guarantee will give you eggs in abundance and at all seasons. It is easy to feed, for all you have to do is to mix it dry in a big box and dip up half a bucket, once or twice a week and fill a box or hopper full of it as the need is. It is quite dry and will keep any length of time.

Formula for Balanced Ration

Mix by measure two parts bran, one part corn-meal, one part oat-meal, one part alfalfa meal, one part beef scraps. Keep some of this in a box or hopper or bucket—dry, perfectly dry—always before the hens. This dry food in the hopper lasts quite a long time, for the hens prefer the table scraps which are fed to them only once a day (at night) and they like lawn clippings, but this dry feed keeps them in just the right condition for egg production —neither too fat nor too thin.

If you do not want to take the trouble to mix this for yourself, you can go to any of the poultry supply houses and buy the food already mixed. This food when put up by reliable firms is what is called the "balanced ration"—that is, it contains the elements of the egg—and when the hens are fed this they simply cannot help laying. They are egg machines which turn the properly balanced ration into eggs.

THE MOULTING SEASON

The moult with hens in the natural state lasts from sixty to a hundred days, but with some hens, especially with hens that have hard, close-growing feathers, the moult and the results of it will sometimes last over a hundred and fifty days; in fact I have known of some that went six months without laying any eggs. Too long to spend half a year dressmaking. Think of the loss to their owners! I did not wonder at the man who told me of it, saying that he just turned them out and "let the blamed things rustle for themselves," but I thought if he had helped them "rustle" perhaps they would not have been so long about it.

Let us consult Nature, as you know I am very fond of doing. After the wild bird has raised her young and her responsibilities are somewhat over, she moults. The older she is the longer and slower is the process of dropping her feathers and growing them again, because as she ages her vitality is gradually lessening. It is the same with hens; the older a hen becomes the longer will be the period of the moult, and not only that but the later will it commence. Let us again turn to Nature and in this copy her. We want the old hens, if we keep them at all, to be the parents of our young next spring and we are only keeping them over for a certain reason (or for sentiment), as they have, perhaps, proved themselves to be our very best layers, or as the parents of our prize winners, or may be prize winners themselves and therefore we want their offspring in the hopes of perpetuating these excellent traits.

The Starving Process

How shall we help these elderly hens to get quickly through the moult? Some years ago I read of a man in New York State, who claimed he could make his hens moult at any time of the year and therefore he could also, by controlling the moult, make his hens lay at any time of the year. His plan was to starve the hens and so stop their laying, and when they had stopped for a week or two he fed them highly with fattening food. This he said made them moult and drop their feathers very quickly, so that in a few days the hens would be almost nude and the new feathers would come in very rapidly. His theory was that when hens sit for three weeks on eggs and raise a brood of chickens they moult quickly because they grow thin during incubation, and when they have the rich feed which is given to the little chicks, it makes them shed their feathers and assists the moult.

His theory sounded very plausible and I decided if he could do it I could also, and tried. I discovered the New Yorker was only partly right in his deductions and that it does not pay to force Nature out of season.

The following year I was much more successful, for I only attempted to "assist" Nature and not to "force" her. I did not try to make the hens moult in June, but waited till nearer to the natural time of the moult, that is, until August. I then put the hens on green food. I know that is hard to get at that time, but I had lawn clippings, vegetables and melons, or even alfalfa hay cut in the clover cutter and soaked for some hours in water, and I dispensed with all the grain and meat. I kept them on this green food for about three weeks until their avoirdupois was considerably lower and most of them had stopped laying for a week.

Dipping Fowls.

Meanwhile during their fast I saw that they were entirely clean from lice, either by keeping them well dusted with insect powder or by giving them a good, warm bath in warm soap suds, rinsing them in a two per cent solution of carbolic acid or water and creolin or the kerosene emulsion. I have tried all of these with good success.

This washing seems to loosen the feathers and will clean the fowls of lice. If lice are left on the fowls at moulting time they eat little holes in the tender sprouting feathers and these little holes in the web of the feather will certainly bring a "cut" from the judge in the show room, and for the whole year will tell the tale of careless handling by the owners. In washing or dipping fowls for lice there are two things to be remembered: First, do it on a bright, warm, sunny morning, so the fowls will have time to get thoroughly dry before sundown, and, secondly, see that every feather is thoroughly soaked. If you skip a feather a louse will take refuge on it and commence to breed again as soon as the hen is dry. If there are any lice the disinfectant in the bath will kill them and the warm suds also loosens the nits of the head lice. Those lice lay two silvery, white nits at the shaft of the feather and it is difficult to get them off.

Mature hens which are fed sparingly for about two weeks and then receive a rich nitrogenous ration, moult more rapidly and with more uniformity and enter the cold weather of winter in better condition than the fowls fed continuously during the moulting period on an egg-producing ration.

What to Feed

It is largely a question of what not to feed as well as how little to give the birds you wish to moult early. There is one line of foods that you may feed in unlimited quantities, and that is the green vegetable, the waste, small beets and thinnings of the garden rows can be supplied every day. My own plan in the days when I had small ungrassed yards, was to give full quantities of lawn clippings, putting them into the yards an hour before dark. This gave the birds time to fill up at night and yet the uneaten clippings would be still fresh in the early morning. If you have had no experience in the use of lawn grass you will be surprised to see how much a few hens will eat. If your hens have very large yards, with fruit trees to supply some falling apples or pears, the birds will do very well without other food. We are inclined to overfeed our birds with grain in the warm weather and, unless the food is really much less than usual, you will fail in getting an early moult.

This low feeding or starving process as it is called by many, is the important factor in the forced moult. Unless you really do this in good shape the birds will continue to lay and will shed their feathers in mid-autumn.

Handle your birds on the roost to test their weight. They must be thin in body, yet good in color of comb and wattles. I find that birds take from fourteen to twenty-one days to get real thin. You will notice as you put this plan into practice that the egg yield will drop off until no eggs are being laid; that the birds are on the run all the day long, coming to meet you at any point of the fence you may approach. The birds show that they miss some of their usual food. This thinning will do no harm to the birds; in fact it adds to the health of the birds for months to come.

The Full Ration

When the birds have lost all superfluous flesh, when the eggs have ceased to appear for a week, feed them good, full rations of growing foods. Now is when you add meat, beef scraps, green bone, cornmeal, and linseed meal. You can give them a morning meal of two parts cornmeal, three parts bran, one part beef scrap. At noon feed a small handful of wheat or barley to every bird and at night a full feed of wheat or corn. Do not neglect to furnish full supplies of green food and vegetables all the fall.

The change from the low feed to the full rations will be followed by the rapid dropping of feathers. The feathers will fall off all over the birds so that many of them will be almost naked. This result will be seen in most of the birds. A few will fail to respond, more if you do not follow the plan as outlined.

Keep the full feed up until the birds get the new coat of feathers and begin to lay a few eggs. Then feed them as you do the fully mature pullets; avoid feeding of heating foods (corn and corn products) lest you start another moult in the late autumn.

The forced moult is ONLY FOR MATURED FOWLS, or fowls that are over a year old. You must not starve the pullets. You must keep them growing. They will stand more heating food than hens. Let the pullets do most of your winter laying, but do not neglect anything that will induce the older birds to give you a good share in the profits of winter eggs.

To sum up the whole matter in a few words, if you want to hasten the moult, do not try the experiment with all your fowls, but take a few, separate them from the others, and about the middle or end of August commence to shorten their food. You can do this suddenly, giving them only green food and all the green feed they want. Secondly, keep this green feed up for two or three weeks, or at least one week after they have stopped laying. Thirdly, the green food should be clover, lawn clippings, alfalfa hay cut in a clover-cutter and soaked in water; beet tops, cabbage, lettuce, etc. Fourthly, after the three weeks' fast, feed rich food, fattening food, sunflower seeds, kaffir corn, wheat, barley, oats and meat. Fifthly, when they begin to lay on this food, which they will do in about a month when they have completed their coat, gradually change the food, taking away the corn and its products, and the linseed meal, and anything that would be very fattening.

Color of Feathers and Skin

The feeding of the fattening foods adds heat to the body, fever our grandmothers called it, and this fever seems to loosen the feathers all at once—just what we want—and they fall so quickly that the hens are almost nude. Then is the time for care in feeding if you have exhibition stock, for I am certain color can be greatly controlled by food.

Now, I know by my own experience that yellow corn will give yellow feathers (brassy feathers) to white fowls when freely fed; that cottonseed meal will have the same effect, for that is what we add to the fattening food the last week to give the yellow tint to the skin. I know that iron in the drinking water has the same effect with white fowls. With colored fowls, such as Brown Leghorns or Partridge fowls or Buffs the iron and the corn will intensify and make more brilliant and bright their colors.

The fowls that are making their new coats, the coats that have to last the hens a year, all need plenty of green food and grain. The white fowls instead of yellow corn, should have oats, hulled oats are best, but if you cannot get hulled oats, soak the oats in scalding water so the hulls will be softened. Hulled oats may appear to be more expensive than the unhulled, but there is so much waste, so much indigestible fiber to the unhulled oats, that I decided that it was more profitable to feed the hulled oats. For those who are feeding cockerels which they want to exhibit in the winter; for the white or black and white, give them shade, plenty of shade, for our California sun will draw out the yellow; cut off all the yellow corn and all cottonseed meal; feed oats, wheat, barley, grit, charcoal and have granulated bone always before them. For the colored fowls add linseed meal to the ration. It will deepen and brighten the colors.

FINE HEAD BUFF WYANDOTTE. O. S. HOFFMAN.

VALUE OF ECONOMY

The old saying, "a penny saved is a penny earned," may well apply to the poultry business. To make money in the business, one must practice economy in every direction.

Economy in Grain

First: Economy in buying the food. This is very important. The available grains vary in different places in price; in some localities, for instance, barley is cheaper than wheat, then utilize barley; that is to say if there is a decided difference in the cost, remembering that barley has a husk on it, which is indigestible fiber, and that fowls do not like it as well as wheat, although they eat it readily if rolled, or soaked or sprouted, and the analysis shows the same nutritive ratio as wheat. Again in some places, oats can be obtained very cheaply, and this is a most valuable grain for feeding and building up large, sturdy frames in the young fowls, promoting egg laying and inducing fertility in the eggs. I have great faith in oats—it is good for man, beast or bird, but the husk is the difficulty there. The oats should be scalded or clipped, or better still, hulled, to make it thoroughly available. In Oregon and Washington, oats is less expensive than in the south, and therefore should be freely used there. By commencing the use of it early, the chicks will be vigorous and of large frame.

Then, again, rice, rice hulls and rice bran are cheap in certain localities, such as in San Francisco and Seattle, where large quantities are imported and cleaned, and these can be had very cheaply and utilized either in the dry or wet mash. In other places where beans and peas are grown in quantities, the refuse of these, which is not worth marketing, can be used most advantageously.

Broom corn seed is a most excellent food and costs very little. I had in Oklahoma many tons of this, to which the fowls had free access and with green growing winter wheat, a little milk and table scraps, they laid all through the moult and through the winter, notwithstanding the blizzards and zero weather. Nothing seemed to stop their laying, and I attributed it to the broom corn seed. Sorghum seed is equally good.

Another little economy I found quite good among the little chickens was buying dry or stale bread from the bakeries at 25c a sack weighing 25 pounds. This I took home, cut same in slices and dried in the sun or in the oven, ground in the grist mill and used either moistened or dry, for chickens, turkeys and ducks.

Economy in Vegetables

Then, again, there are the various vegetables, many of which can be had for almost nothing. There are "small potatoes." It generally raises a smile to talk of these, but they make a most excellent addition and variety to the fowls' bill of fare. Small raw potatoes can be chopped up in the chopping bowl in a few minutes, also turnips, carrots and onions, and the outer leaves of cabbage, cauliflower or celery. I bought the largest chopping, or butter bowl, I could find, and a double-bladed chopping knife, and used it every day, especially for the little chickens and turkeys. Small potatoes, turnips and carrots can be boiled, mashed, mixed with bran and blood meal, or with milk, and make a good variety in the diet. If you have other vegetables to spare such as beets, cucumbers, pumpkins, etc., and find the fowls do not at first like them, chop some up and mix bran with them and soon the hens will acquire a liking for them.

Another economy is using the leaves which fall from alfalfa hay. When the haymow begins to get empty, sweep up the leaves and put them in a box or sack to mix in either the dry or wet mash. I used to try to keep the last two bales of the alfalfa hay, as the balers would sweep up the leaves and put them in these last two and this was just what I wanted for my hens. Sometimes I soaked the leaves and fed them at noon, keeping the alfalfa tea to mix in the mash with potatoes and bran or whatever I was feeding. I always said the alfalfa tea was as good as beef tea. There are many ways of economizing in the feed.

Eocnomy in Labor

Another thing to economize is labor. I know many a farmer's busy wife will agree with me in this. I found the dry feed a great saving of time and strength. It was much less labor to carry around to my many pens of fowls, buckets full of dry food nicely mixed in the proper proportions and pour it into a box, or trough or hopper and let the hens eat it dry, instead of laboriously mixing it with water. Before trying the dry feed, I had so many hens that I had a large trough made, like a plasterer's trough, and I used to mix and turn the mash with a spade or hoe and then fill those large buckets full and put them on a child's express wagon to pull out to the pens. This was quite hard work and I hailed with joy the easier task of carrying the lighter buckets of dry food. I found, too, that it saved time to mix up the food by the sackful or binful; then all that was required was to dip up a bucketful for each pen. I showed this plan to a friend of mine and later had a letter from her telling me it was a great comfort, for all she had to do was to send her Jap boy out to that certain box or bin and tell him to feed that; she knew he could not make a mistake for it was ready mixed.

Economy in Water

Another economy: Have a water faucet in each pen. This may seem like an expense at first, but it will pay in the end, for fresh water is as important as good food, and if it requires but a turn of the faucet the hens are sure to be amply supplied. At one ranch where there was an abundance of water, I saw a small fountain which ran into a basin and that in turn overflowed into some cobble-stones and a drain, so that the hens had always fresh water without drawing on either the strength or time of their owner.

I would, however, caution chicken raisers against allowing the water to run in a stream from pen to pen, as that may carry infection, especially the infection of colds and roup. One gentleman who had 3000 fowls told me that letting the water run in a small stream through his pens, had ruined him in the chicken business. One pen at the top of the hill got roup, and the infection was carried through to all of them. In Kansas one of the worst outbreaks of chicken cholera came from a creek. All the farms on that creek lost all, or nearly all, their chickens, from drinking contaminated water. A faucet in every yard would be cheaper in the end than an outbreak of roup or cholera.

Economy in Fencing

Economy in fencing came in very handily one summer. I found I could make a very good temporary chicken-wire fence with posts 50 feet apart by "darning" in a lath every eight feet or so, passing this lath in and out of the wire meshes before putting up the wire. This keeps the wire stretched and when taken down it can simply be rolled up and used over and over again, keeping the lath in it ready for the next time. I found chicken-wire and lath quite an economy. I made cat and hawk-proof little pens of this. Bought a bundle of six-foot lath, some two-foot chicken-wire and made most useful little panels six feet long with the laths, stretching the chicken wire on them and tacking it down with two-pointed tacks. I wired or tied the panels at the corners and had a larger panel go over the top made of six-foot wire. I did not have to kill any cats or have fusses with the neighbors. The little panels were untied and piled up for the winter time and put in the barn, coming out almost as good as new the next season. They were cheap, light, easily handled and very satisfactory.

Beware of Spoiled Food

It is poor economy to buy spoiled grain of any kind. The best is none too good, and anything that is spoiled is very apt to bring in diseases. Wheat or any grain that has been moistened will develop fungoid growth; smutty wheat, etc., is almost poisonous to fowls, while, of course, we know that there is no grain that so nearly approaches the analysis of an egg as does wheat, when it is good. Corn, likewise, if it has been dampened, will commence to ferment and that will disagree with fowls. At one time there was a fire at a flour mill in Los Angeles. A great deal of the spoiled wheat was sold for chicken feed. "Anything is good enough for chickens," was the cry, and hundreds of chickens lost their lives from that wheat. The owners of the fowls thought it was chemicals that had been used in suppressing the fire, but it was nothing but water, some of the firemen told me, that had been used for extinguishing the fire. The dampened wheat became musty and mouldy and it was that which killed the chickens. Again in using beef scraps, meat meal, blood meal or animal meal, be careful to buy the best you can get, and keep it carefully away from any dampness. Dampened or spoilt animal food is poisonous to the chickens and many a fowl has died from ptomaine poisoning from using spoiled animal food. One of the greatest economies is to buy in large quantities.

Most Suitable Green Foods

Whilst we are on the subject of economy we must not forget the two green foods that are the most suitable for fowls—clover and alfalfa.

Let those who are living on a town lot have a clover lawn; clover requires less water than blue-grass or any lawn grass in this climate, and is easily grown when once it is properly started. The lawn clippings are just the right length for green food and if necessary, the hens can be turned out on to the lawn two hours before sunset, and will then busy themselves nipping off the clover leaves; they will not have time or inclination to do damage by scratching. A run on the lawn before bedtime is a wonderful tonic for chickens that are yarded closely all the day.

Every farm should have an alfalfa patch, if not a good big field of alfalfa, and no chicken ranch is complete without one, for the youngsters should have a good alfalfa run to properly develop them.

Alfalfa is a legume; is rich in nitrogen and enriches the land upon which it is grown. It is the best green feed next to clover for the hens or cows, and the hens love it. It is equally good for ducks and turkeys. The question of economy of labor is a very serious matter in poultry raising, and by having a good alfalfa patch upon which the hens may be turned several hours daily, the labor of cutting and preparing green food for them is eliminated and will prove a great economy.

Hens that have an abundance of alfalfa will lay eggs with very rich colored yolks and these eggs are usually fertile and produce healthy, vigorous offspring. An alfalfa range insures health, a good digestion and to growing chicks, a large frame. In buying a chicken ranch, one of the important questions is "will the land grow alfalfa?" Is there sufficient water to raise a good crop of alfalfa?

Alfalfa meal, or as it is sometimes called, Calfalfa, has been successfully used for hens. This is alfalfa hay ground up finely to form a meal. I have used this for several years and I find it sometimes good and sometimes bad. The analysis of it made by the University of California shows the protein content to be very high, and the nutritive ratio to be 1:3.3. This is the good meal. The poor meal contains too much fiber, and, as Prof. Rice of Cornell University remarked, "It is better for stuffing a bed than a hen." It all depends upon the quality of the alfalfa. Sometimes it is left until it is too old or is not properly cured, and is almost valueless; at other times it may have been dampened and become musty. When this is the case, it will disagree with the fowls and give them diarrhoea. To test it, pour boiling water upon it, and if it smells sweet, like hay, it is all right. If there is a musty, mildewy smell, discard it.

PRESERVING EGGS

Of twenty methods of preserving eggs tested in Germany, the three which proved the most effective were coating the eggs with vaseline, preserving them in lime water, and preserving them in water-glass. The conclusion was reached that the last was preferable, because varnishing the eggs with vaseline takes considerable time and treating them with the lime water may give them a disagreeable taste. These drawbacks are not to be found with eggs preserved in water-glass, which unquestionably is the best preservative yet discovered. The most difficult point probably in the use of water-glass for preserving eggs is its tendency to vary in quality. As a matter of fact, there are two or three kinds of water-glass, and in addition to the fact that the buyer does not always have a distinct idea as to what he wants, the local druggist may not know all about it, or he may not know which kind is best for preservative purposes. The main use of these preparations for years has been the rendering of fabric non-inflammable. This use in the Royal Theater of Munich has rendered the place fireproof by its use as a varnish in the fresco work, woodwork, scenery and curtains. It is also used for hardening stone and protecting it from the action of the weather. It was thus used many years ago, to arrest the decay of the stones in the British Houses of Parliament. The use of this medium for egg preservation is comparatively new, especially in this country, and it is not to be wondered at that dealers do not always supply just what is wanted.

Different Names for Water Glass

If we used the term soluble glass or "dissolved glass" in preference to either water-glass or silicate of soda, it might better describe just what we want, although one of the other names might be preferable when ordering of the druggist. This term expresses exactly what the material is. When we buy it by the pint or quart, we get dissolved glass. When we buy it dry, we get a soluble glass powder sometimes like powdered stone, sometimes white and glassy as to its particles. The powdered forms are supposed to dissolve in boiling water, but they do not dissolve readily, and must often be kept boiling for some hours.

Water-glass is made by melting together pure quartz and a caustic alkali, soda or potash, and sometimes a little charcoal.

Several of our Experiment Stations have made some rather exhaustive experiments with this dissolved glass in preserving eggs. The reports are, without exception, in favor of it. No other preservative is reported as being equal to this one. The stuff is invariably described as a thick or jelly-like liquid, and the proportions recommended are one pint of the silicate of soda to nine pints of water, although the Rhode Island Station reports experiments in which as low as two per cent of water-glass was used with favorable results. This is done to find out how little could be used, but this small proportion was not recommended. Further trials may show that less than nine to one may be reliable.

Directions for Use

The directions for use are: Use pure water which has been thoroughly boiled and cooled. To each nine quarts of this water add one quart of water-glass. Pack the eggs in the jar and pour the solution over them. The solution may be prepared, placed in the jar and fresh eggs added from time to time until the jar is filled, but care must be used to keep fully two inches of water-glass solution to cover the eggs. Keep the eggs in a cool place and the jar covered to prevent evaporation. A cool cellar is a good place in which to keep the eggs.

If the eggs be kept in a too warm place the silicate will be deposited and the eggs will not be properly protected. Do not wash the eggs before packing, for by so doing you will injure their keeping qualities. Probably by dissolving the mucilaginous coating on the outside of the shell. For packing use only perfectly fresh eggs, for eggs that have already become stale cannot be preserved by this or any other method, and one stale egg may spoil the whole batch.

I can speak from my own experience, for I have packed eggs in it for five years and shall do so again. We are fond of fresh eggs and use a great many, and I find it most convenient to have a jar or crock full of nice eggs always on hand. I have kept them myself for eight months and have no doubt but that I could have preserved them still longer had we not eaten them, for I found them to all appearances as fresh as if not over a week old. It cost about 1½ cents per dozen to preserve them.

The Kind of Vessels for Packing

Prof. Ladd, of the North Dakota Agricultural Station, spoke of receiving a few complaints that barrels were not proving satisfactory, the water-glass appearing to dissolve some product which deposited on the eggs. He thinks this might be attributed to the presence of glue, which had been used as a sizing for the barrels. In such instances, charring the barrel inside with thorough washing thereafter, is recommended. Altogether, the preference seems to be for glass or stoneware vessels.

Prof. Ladd's statement as to the satisfactory results of the water-glass method is very strong. He says: "This method has been tested in a commercial way, in nearly every state and part of our country, and we have not had to exceed eight adverse reports." One of the stations affirms that the failures reported are probably due to receiving water-glass of poor quality.

It is also stated that these, like all preserved eggs, contain a little gas, and, when boiled, they will be likely to burst unless previously pricked through the shell at the large end.

As the entire processes of preservation are an effort to fence out germs, the recommendation not to wash off the mucilaginous coating which nature puts on the eggs, and also to use only boiled water, appear very logical. When we know just what we are aiming at, we are less likely to omit the little precautions which otherwise might seem like the whims of some fussy person. Too many people skip the essentials when trying to follow a formula.

I have kept the eggs in tin receptacles, five-gallon kerosene oil cans and large lard pails. These kept the eggs perfectly, but after a time the water and silicate of soda rusted them in spots and the red rust formed a sediment on the eggs. This did not injure them as far as I could see, except giving them a brownish tinge, and on asking the druggist, he said he did not see why the tin should not be used, as the silicate of soda comes from the East in tin cans. If tin is used, it is best not to paint the cans or oil them, as the soda has an affinity for oil and will eat through it and the oil or grease may impart a disagreeable flavor to the eggs. Remember the eggs must be absolutely fresh, for one bad egg may spoil the whole quantity in the receptacle.

Preserving in Lime

The process of keeping them in lime-water is as follows: Slack four pounds of lime, then add four pounds of salt; add eight gallons of water. Stir and leave to settle. The next day stir again. After the mixture has settled the second time, draw off the clear liquid. Take two ounces each of baking soda, cream of tartar, salt petre, and a little alum. Pulverize and mix; dissolve in two quarts of boiling water. Add this to the lime water. Put the eggs in a stone jar, small end down, one layer on top of another, and pour on the solution. Set the jar away in a cool place. This method is quite satisfactory, but not so good as the water-glass, as the eggs are liable to taste of the lime.

ROSE COMB WHITE LEGHORN. COURTESY RURAL CALIFORNIAN.

CAPONS

"Does Caponizing Pay?" We will consider the matter fully and from different points of view.

In Philadelphia and New York, in London and Paris, capons are considered a great delicacy, and as we, in California, become more metropolitan, capons will be more and more in demand. Eleven or twelve years ago when I had capons for sale I could not get more per pound for them than for the uncaponized fowls, as the Angelenos had not been educated in taste to the excellency of capon meat.

Capons are undoubtedly a more delicious dish at a year old than an uncaponized male bird of the same age. I had been led to suppose that a capon would be immensely heavier and larger than an uncaponized bird of the same age. This I found was not the case, the capons being rarely more than from half a pound to a pound heavier, if at all. My chief reason for caponizing was the desire to train capons for foster mothers of chicks. I wanted mothers that would not commence to lay as my hens did when chickens were two, or at most, three weeks old and then desert them. In this I was thoroughly successful. The trained capon will mother chicks just as long as the chicks will stay with him, and after a little rest will take another brood and mother it again, clucking to the chicks, feeding them, defending them, hovering them better than the hen.

"Does caponizing pay?" Careful experiments have proved that the increase in weight is by no means so great as the public has been led to believe. It takes capons at least a month to sufficiently recover from the operation to catch up with their former mates in size and when they come to a marketable age they seldom weigh a pound more than the uncaponized birds of the same breed and age. The gain, however, in price is in their favor, for it about doubles that of the other. This sounds like a strong argument on the side of the capon, but again the cost of production is an essential factor in the study of the question. It will cost as much to produce a ten-pound capon as to produce three or four young chicks of the same combined weight; in fact with food at the present price I really think it will cost more.

"Does caponizing pay?" I knew a lady about three years ago who sold four capons for sixteen dollars. She was so much encouraged by this, for they averaged 38 cents a pound, that the following season she drove around the country buying up little cockerels and caponizing them. She was very successful in operating, rarely losing any, but as she only stayed in the business one year, I think she did not consider it very remunerative.

Easy to Learn

The art of caponizing is simple and easy to learn. In France the farmers' wives and daughters have done the caponizing for centuries and practically without instruments except a sharp knife.. In this country and age, we can buy a case of the best instruments, with full instructions for use, at a low cost, and the agricultural stations of some states give free demonstration lessons to anyone within the state. The Rhode Island College gives lessons in caponizing in connection with its poultry course and also sends out, free, a book of instructions. By following these instructions and experimenting for the first time on a dead chicken, anyone that is deft can learn it. The operation is performed with apparently little pain to the subject, and the minute the bird is released it will eat heartily and walk around as if nothing had occurred.

In foreign countries the art of caponizing has been known and practiced for ages, yet it is not so common nor are capons so plentiful but that prices rule high and capons are considered the choicest of viands and above the reach of any except the rich. In this blessed country there is no reason why the producers of poultry should not feast upon capons, besides having the satisfaction of producing and marketing strictly high-class poultry.

Favorite Breeds for Capons

In New England the favorite breeds for caponizing are the Light Brahmas and the Cochin and Brahma crosses. They are chosen on account of their large size and slow growth to maturity. The Plymouth Rocks follow, together with the Orpingtons and Wyandottes. The smaller breeds make, of course, much smaller capons, still they are popular in small families where large size is not required. I have personally caponized only my White Plymouth Rocks. Nothing could be better than capons of this breed. At nine or ten months of age they are in their prime and the juiciness and flavor of their flesh is superb.

Among the advantages of caponizing are, the birds may be kept together in large numbers, will not quarrel or fight, will not harass the hens and pullets, will not misuse the little chicks, bear crowding and take on flesh more rapidly than cockerels. They make, when trained, most excellent mothers for little chickens, sheltering them under their long feathers and great wings.

Best Time for Caponizing

The best time for caponizing is in the early fall, for the reason that the heat of summer does not then retard recovery and also because the late (June hatched) cockerels are then of the best size.

The best size is from two and a half to three pounds weight and this would be about the weight of June hatched chickens of the American breeds which if caponized in September will be well grown and in good shape for marketing in March, the time of the highest prices.

It is to the farmers, however, that the recommendation to caponize their cockerels for the family table should appeal most strongly, for they are the class that would be most benefited by having good capons to eat. It is a simple task to caponize forty or fifty birds and by that simple method a farmer can provide his family with dinners which will be the envy of his less fortunate friends.

The question, "Does caponizing pay?" may be answered, "Sometimes it does and sometimes it does not."

Capons as Brooders

Capons make excellent mothers when trained to it. Some breeds would probably make more affectionate and attentive foster mothers than others. I can personally answer for the Cornish Indian Games and Plymouth Rocks. I have also seen beautiful Brown Leghorn capons that had raised several broods of chickens. Cockerels hatched in November, December and January, make excellent capons for brooding. They should be caponized at about three months of age. Should be gently handled and never frightened, when they will become perfectly tame. The capon with its changed nature is even more timid than a hen or pullet, and for this reason should be separated from any of the older fowls and kindly treated.

Capons should be trained at the age of about six months. They are easier to train at this age than at any other time, generally, but I have trained them at ten months of age. To train them, I keep the bird in solitary confinement for a few days, placing him in a cracker box; place water, grit and sand in the box the same as though preparing for a hen and her brood. After two or three solitary nights and days I put two little chicks under him at night; they snuggle up under him, and he is quite glad to have the little fellows for company. The next morning he will look a little surprised perhaps, but usually takes them immediately, and soon begins to cluck to them like an old hen. The following evening I put as many as I intend him to care for under him, and before going to bed at night see that all the little fellows are under his sheltering feathers. My object in using the cracker box is that it is about the proper height to make it uncomfortable for the capon to stand upright and he will sit for comfort; the little chicks get closer and make friends quicker, and have an opportunity to nestle under the capon as they would a hen. This training should be done in pleasant weather, because the chicks will not be hovered at first as well by the capon as the hen, and I use only a few chicks the first time, because a young capon with his first brood does not hover them like a trained one.

The Whiskey Treatment

Hen-hatched chicks take to a capon without any trouble, but chicks which have been several days in a brooder seem afraid of the capon, and instead of running to him to be hovered, huddle in a corner, so it is best to put them straight from the incubator under the capon. A writer on this subject says: "Should one of the capons pick the chicks I would take him out of the box and swing him around in a verticle circle at arm's length until he was sick, then put him back again. If he attempts the same thing again, I take a small glass syringe and inject about one tablespoonful of good whiskey into his crop through his mouth, and after this treatment he is pretty sure to take to the chicks. He becomes so docile that he allows the chicks to pick at his face and will not pick back at them. When you notice this, you can rest assured that he is on the right road."

I have never tried the whiskey treatment, and have never had any difficulty in training a capon. Capons have proved far superior to hens in brooding chicks, in fact they excel all other methods, either natural or artificial. The hen, especially "bred-to-lay" strain, deserts her brood at too early an age, and some hens, especially the pullets with a first brood, are often very stupid at caring for them. I have known a pullet to hover her chicks in a thunder storm in a gully where the water rushed until they were nearly all drowned. Pullets do not seem to have sense enough to "come in out of the rain," while a good capon, when once he has been taught his way home, will bring the little ones to shelter without any trouble. The capon will defend his little brood most vigorously against cats, dogs or any animal. He seems to develop all the latent parental affection and lavishes it on his young charges as if his one and only object in life was to care for them.

When Changing Broods

When the chicks are old enough to take care of themselves, before entrusting another brood to his care, he should have a rest of at least two weeks, especially if the next brood is to be of another color. During the two weeks' rest he will forget the color of the chicks he had and will not be so apt to object to the new ones. We all know that hens will sometimes object to chicks of a different color and will oftentimes kill them. When once trained, a capon is very little trouble and will care for brood after brood without any more training than I have mentioned. Capons can be kept over several seasons. I have heard of some being used for eight years, but mine were usually fattened and made a toothsome dish after two years' service.

It is not difficult to learn how to caponize. The tools or instruments necessary are to be found at the poultry supply houses. The price for a set of instruments is from $2.50 to about $4.00, largely depending upon the case in which they are contained. The poultry supply houses have books of instruction for caponizing, and at some of them you can learn the names of persons who, for a small sum, will caponize for others. It would be a good plan for several neighbors to join together and have the person caponize 50 or 100 in the same day. In this way it would make the price lower.

Capons are not much larger than cockerels of the same breed and age. The difference is in the table quality of the flesh. It is juicier and more tender, just as steer beef is superior to any other beef.

TURKEYS AND HOW TO RAISE THEM

Turkeys.—The turkey is the largest of our domestic fowls, and the only one that can be credited to America. To get its early history, as far as civilized countries are concerned, it is necessary to go back to European records. Until a few years ago it was thought that its first introduction into Europe occurred about 1524-25, but in Brown's "Races of Domestic Fowls" there is a supplementary note saying that the librarian of the Royal Palace, Madrid, recently (in 1906) discovered that it was introduced into Spain as early as 1500, eight years after the discovery of America. Pedro Nino, a Spanish adventurer, discovered this fowl on the coast of Cumana, north of Venezuela, in 1499, and the next year took it to Spain, along with such other curiosities as parrots, monkeys and bright-colored birds of other kinds. Turkeys were bred in great numbers by the Aztecs and other nations of Mexico and Central America. The bird was not found further south than the mouth of the Orinoco on the east coast, but on the west coast it was found as far south as Peru.

How this bird got its name has been a matter of curious speculation and much traditionary lore has been put forward as authentic history. As a matter of fact not a single one of the old writers attributes its origin to Turkey, all agreeing that it came from America. Wright thinks the name might have been given it from the red head and neck. Others suppose that the name comes from the domineering habits of the bird. Unfortunate as the name was, it has been accepted and speculations as to its origin are of no value. The turkey was brought into England in 1521 or 1524. In 1541 it was so highly esteemed that the Archbishop Cranmer prohibited the serving of more than one turkey cock at state dinners and the serving of turkey hens was forbidden altogether, as they were too valuable to eat. About 1570 the turkey became the recognized Christmas dish with the English farmer.

It is said that the first turkey was eaten in France at the wedding of Charles XII and Elizabeth of Austria, June 27, 1576. A large number of the birds had been sent over from Boston to St. Malo and when the ship reached that port the provincial governor sent a dozen of them to the king's cook. The king was so pleased with them that he began to breed them, and the breed rapidly spread over France.

Bronze Turkeys.—This variety as bred today is the direct descendant of the wild fowl, although seemingly it has decreased somewhat in size, as wild ones weighing 60 pounds are recorded. There were three varieties of turkeys in America at the time of the discovery: the wild turkey from which our bronze variety comes, a smaller Mexican variety, and still further south a variety which lacked the tuft of hair on the breast but has a crest or aigrette consisting of a single row of feathers on the head. The latter is indigenous to South America and is called domesticated, but does not endure removal to colder climates.

SOME OF ED HART'S MAMMOTH BRONZE TURKEYS ON THE RANGE, CLEMENTS.

The bronze turkey is the heaviest of all the varieties. It requires two years to get the full weight, at which time it should weigh 36 pounds for cocks and 20 pounds for hens.

Turkeys have been called the "farmers' friend," and there is no doubt that turkey raising on a small scale is more profitable than any other branch of the poultry industry and that turkeys will bring larger cash returns than any other stock upon the farm. They cost very little to raise, they eat the waste grain in the fields and barnyard, besides the seed of many harmful weeds. They consume an immense number of grasshoppers, grubs, worms and insects which would otherwise greatly injure the farmers' crops, and they are not difficult to raise if they are not overfed.

One writer asks if chick feed is a proper and safe food for little turkeys, and another requests me to tell her exactly how I feed and care for the little turkeys.

Chick food is neither a safe nor a proper food for little turkeys, although it is a most excellent food for little chicks. In fact, you may be sure of success when you feed it to chickens and failure if you feed it to turkeys. Later on I will try to explain this.

Now, as to my way of rearing turkeys. I am glad to give it, because now I raise every turkey that is hatched, barring accidents, as some will drown in the cows' trough and occasionally one or two get stepped on, or the door blows on one, or the puppy worries another. None die from disease.

I do not pretend to say that mine is the only way, but I do say that not only do I succeed in raising turkeys, but those who have followed my directions were as successful as I have been, and those that met with failure did not follow my plans. I have been criticized as too fussy and particular about little details, but I think it pays to take good care of the little things for a few weeks, for turkeys are delicate only when they are little, and if properly cared for they will be strong and hardy when they mature.

Grandmother's Recipe

At my grandmother's the recipe for feeding little turkeys was as follows: "Leave them in the nest twenty-four hours or until the mother turkey brings them off; then give them only coarse sand, and water to drink. Meanwhile put some fresh eggs in cold water to boil; let them boil for half an hour; then chop them up, eggshells and all, quite fine; add an equal amount of dry bread crumbs, and always, always, some green food chopped up finely."

Lettuce, dandelion or dock were the green foods at grandmother's, and the explanation given me was that if they are fed without having green at every meal, they soon become constipated, then get sick and die. The secret of her success was the tender green food and the grit, a pinch of coarse sand being sprinkled over the food of each meal. As the little turkeys grew, a little cracked wheat and later whole wheat was added to their food. That was the only grain given. This was grandmother's recipe for raising turkeys.

The way I feed and have fed for years is as follows: When the little turkeys are twenty-four hours old I put freshly-laid eggs into cold water and boil them for half an hour; chop them up fine, shell and all; add equal parts of bread crumbs; feed dry, taking away what they leave, feeding the mother separately.

The next day I feed the same, adding very finely chopped lettuce or dandelion leaves or green young mustard leaves and tender young onion tops. This is their breakfast and supper. For dinner they have a little curd made from clabber milk, cottage cheese some call it. In a few days I add cracked or whole wheat to their supper, and if I am short of bread crumbs I add rolled breakfast oats to the egg and bread crumbs I always chop up an onion a day with the egg, and bread crumbs unless the onion tops are very young and tender. Onions are an excellent tonic for the liver and kidneys, and prevent worms and cure colds; so I use onions freely both for turkeys and chickens. In a few days I commence to add wheat to their food and at two weeks of age I gradually arrive at giving them wheat and rolled oats for breakfast; in the middle of the forenoon a head of lettuce to tear up and eat; at noon cottage cheese, and about four or five o'clock their supper of egg, bread crumbs or rolled oats, lettuce and always the chopped up onion.

I give them clean water three times a day in a drinking fountain, or if I have not a fountain I make one out of a tomato can. Make a nail hole in the can about half an inch from the top, then fill the can up to the hole with water, invert a saucer over it, and holding the saucer tightly to it, turn it over quickly. This makes a good fountain, for the water will come slowly out of the nail hole into the saucer. I give the turkeys a similar fountain of skim milk, also. A word about the cottage cheese. I am very particular in making it not to allow the clabber milk to become hot. I use either a thermometer, letting the heat only come to 98 degrees, or I keep my finger in the milk, and as soon as it feels pleasantly warm I take the milk off the fire, pour the curd into a cheese cloth bag and leave it to drain. If the milk scalds or boils, the curd will be tough, hard like rubber and indigestible enough to kill turkeys or chickens.

Overfed Little Ones

When I lived in the home of the wild turkey, Oklahoma and Kansas, I learned much about the care of tame turkeys. There "corn is king," but I was cautioned never to give corn to the young turkeys until after they "sport the red." That is, until their heads and wattles become red, which happens at about three months of age. It was said that corn always sours on their stomachs. It was there I heard of a man who brought up his turkeys on nothing but onion tops, curd and grit, and they did well.

One of my experiences in the land of the wild turkey may serve as a warning to others. I had a good old Buff Cochin hen who was mothering a brood of nice little turkeys. She was most assiduous in her care of them; she clucked to them all day; called them up to eat all the time, and it was surprising to see how those little fellows grew, when one after another they began to droop

MAMMOTH BRONZE TURKEY AND YOUNG. ED. HART, CLEMENT, CAL.

and die, till only one was left. The other turkeys under turkey mothers were doing well, so I took the lone little one one night and put him under a mother turkey out in the meadow and saved his life. The old hen had overfed the others. Chicken hens are too anxious to feed the little turkeys. They scratch for them, coax them to eat, and the little turkeys are such greedy, voracious little things that they overeat and in consequence die. I prefer to bring up little turkeys under a turkey hen or even in a brooder, rather than under a chicken hen. The best way of managing a hen is to keep her in a coop, letting the little turkeys run outside or else tie the hen under a tree by her leg. I only feed the little poults three times a day just what they will eat up clean in ten minutes. With a turkey hen I can leave wheat in a trough always accessible, and she will never overfeed the young. The turkey mother will take a few mouthsful herself and then move slowly and deliberately away and her babies will follow her, having only taken one or two grains each. This is more like the nature of the wild turkey, and the nearer to nature one can keep in raising turkeys, the better will be our success.

Nature's wild turkeys are only hatched in the spring when there are grubs and worms in abundance, with plenty of green grass and tender leaves and no grain but what is sprouting, and, above all, Nature never mixes mashes to turn sour and ferment on the little stomachs. The hard-boiled egg and the curd take the place of the bugs and the grubs, for we cannot supply the turkey with anything like the amount of grasshoppers, grubs, worms, larvae of insects which Nature provides in the haunts of the wild turkeys. Another lesson we may learn from Nature's book: Wild turkeys are only to be found where there are springs and streams of pure water, and they never wander away from the water. Give the young turkeys plenty of clean, pure water to drink.

There are two chief causes of mortality in little turkeys—lice and overfeeding. Before giving the little turkeys to the mother to care for, dust them well with "buhach," and continue to do this once a week until they are too large to handle. Look for lice on the head and on the quill feathers of the wing and rub the powder well into them. Lice and overfeeding kill thousands of little turkeys. Overfeeding kills more than lice, and if it does not kill them, it stunts their growth, and unfortunately until they begin to die at about six weeks of age, one scarcely realizes that they have been overfed.

Little turkeys have voracious appetites, and if allowed to do so, will eat too much, and it only takes a few weeks for them to eat themselves into their graves. If they hunt for their food, as the wild turkeys do, they take it leisurely, just what they can easily digest, exercising between each mouthful and just enough is digested and goes into the circulation to keep them healthy. I never feed little turkeys all they want, only what they need, and I always keep them a little hungry.

Keep Liver Healthy

I can tell you just how overfed turkeys will die. First they will walk slowly, lagging behind the others, as if tired, then their wings will droop and they will look sleepy and will not eat, will look at the food as if they wanted it, but were too lazy to pick it up, then diarrhoea will set in, the droppings will become yellow and sometimes green, and death will soon follow. If you hold a postmortem examination, as you should do over everything that dies in the chicken yard, you will find the liver of these little turkeys has yellow or white spots on it, and on cutting into it, you may find that these spots are small ulcers that extend through it. Sometimes these ulcers are quite offensive. This comes from overfeeding, which gives the liver more work than it can do and it breaks down.

The liver is the largest organ in the turkey's body, and it seems to be the most delicate. If you can keep that healthy, you will have healthy turkeys. Onions and dandelion leaves are tonic for the liver and the green food keeps it healthy, whilst the animal food and a small amount of cereal will make the frame of the turkey.

Suppose you should see one little turkey in the brood beginning to walk slowly, what should you do? I will tell you what I would do. I would catch that little turkey and give a Carter's Little Liver Pill and follow this the next day with a little Epsom salts for the whole flock, and cut off some of the grain in the feed. You will probably save the flock, but they may be stunted in their growth, and their liver many months later may break down from being weakened by that first attack of liver trouble.

Chick Feed for Turkeys

Now about the chick feed. It is composed of a number of different grains. Some of these grains are extremely difficult of digestion for turkeys. The chief of these are cracked corn, Kaffir corn, Egyptian corn, sorghum seed, millet, etc. I could scarcely believe this until I had ocular demonstration of it. Then I discovered that cracked corn did not commence to digest in the crop; the gastric juice of the crop does not seem to have any influence on it. It passes through the crop and on through the proventriculus to the gizzard, arriving there hard and not in the least softened or digested, and there it commences to ferment, causing diarrhoea or else passing away without digesting. I am not scientific enough to know the reason for this nor why wheat should be softened in the crop and partly digested before reaching the gizzard, but I know that it is so. They told me in Kansas that corn soured on the turkeys' stomachs, but it does not exactly sour, it ferments—and there is where the trouble comes in.

Sour milk is sour, but this is from lactic acid, and lactic acid seems beneficial to turkeys, whilst the souring of grains, bran, cereals of any kind, or cornmeal is a ferment, and ferments are very injurious to fowls of all kinds, and especially so to turkeys.

Mrs. Charles Jones, the best authority on turkeys in the United States, agrees with me about feeding turkeys. She writes:

"A diet of part corn agrees with chickens, but I have never yet fed corn in any form to young turkeys but that sooner or later they would give up the unequal contest. A little neighbor girl that had a great deal of the care of turkeys said the least little bit of corn meal makes them die. She had learned this by watching them as she fed them."

1100 Gleaning Wheat

It was my privilege to visit a turkey ranch in the San Joaquin Valley some time ago and what I saw there made me wonder that there are so few large turkey ranches in California.

There were over 1100 beautiful turkeys gleaning the wheat over many acres of stubble. These great turkeys had been hatched near the barn in shed-like coops, under turkey hens. They were kept in the yard until about five or six weeks old, when they were driven out with their mothers upon the wheat stubble to rustle for their living, to pick up the wheat that would otherwise be lost. All these turkeys roosted in the open air and to this and the simple life, working for and finding their own living, may be attributed their healthiness.

There are many beautiful valleys in California where turkeys may be grown to great advantage by the hundreds and even thousands, but even on small ranches a few may be kept.

Turkey Lore

With the coming of the fall our thoughts turn turkeyward and letters come to the writer telling of the fine success almost every one has had this year with their turkeys. However, some are also finding that the luck which has been theirs all the summer is now slipping away, and some of those want to know the reason why. Mrs. Chas. Jones, the turkey expert, explains this very plainly in one of her interesting talks. She says, "To understand any branch of poultry culture, one must know their nature and requirements. The turkey is the only bird or animal that has previously existed in a wild state that has been domesticated to the extent of being raised on farms from the Pacific to the Atlantic and from British Columbia to the Gulf of Mexico. They may be raised further north, but I have no authentic account of the fact. The prairie chicken has never been domesticated. Occasionally wild geese and ducks have been raised, but the turkey stands alone as a wild bird domesticated, and because people have not studied into their wild nature and what their diet consisted of in their wild state, they have failed to raise more than a small part that are hatched.

When the turkeys roamed over the forests at the time this country was inhabited by Indians, very little corn was raised. A few beans constituted the grain crop, and as the Indians were too lazy to raise the corn themselves, and put the work on the squaws, who had the young braves in the form of papooses to pack around on their backs, and all the domestic duties to look after and keep their wigwams in order, they did not raise corn enough to glut the market and the turkeys had to look elsewhere for their food than to the corn cribs of the natives. The trees were their roofs to keep the young dry, the leaves beneath them made it dry for them to walk over, and the insects that had found their homes beneath the leaves made their food.

Now we are confronted with a different problem. The forests are about gone, and during a wet season the turkeys must be kept near home for protection from the rain. When they have outgrown their baby feed, what shall we feed them? Corn as a steady diet is out of the question; as they will eat large quantities, their livers

ROOT'S WHITE HOLLAND TURKEYS ON RANGE.

will become congested from such highly concentrated food, and you will notice that one goes a little slow. That will be the only symptom of anything amiss; it will walk a little slower, until it turns up its toes to the blue sky, a silent protest against letting them gorge themselves on the most highly concentrated food that grows.

Wheat, where it can be had, is the best and safest grain that can be fed, but wheat here represents gold dollars and is too expensive to feed. Oats has too many hulls. This year, on account of the high price of wheat, we could not get shorts. Shorts, moistened with water or milk, can be safely fed to turkeys. The turkeys found that we had plenty of corn in the crib and they developed a great liking for it and only wanted to loaf around the crib, until I lost two or three, and then I just turned them out and made them get their living in the fields, and they are all right again. Letting them live on corn is entirely against their way in their wild state, and they have to suffer the penalty. It is like raising children on rich pie, cake and preserves. Their digestive organs get out of order, they become sick and often die.

When turkeys find that they must stay out on the range and get their living they are soon content. By instinct or by personal investigation of the different fields they soon find where the largest number of grasshoppers are to be found. They are methodical in their proceedings, getting off the roosts just at daylight and going to the nearest fields, picking all the insects they can find, then coming to the house for a drink and starting out for the larger range for the day.

At first, when turned out on range, I feed them at night, as an inducement to come home early, but after the habit is established I think it better for their health not to feed them, as they come home with full crops and any more crowded into the digestive organs impairs them.

MORE ABOUT TURKEYS

There is no need for any sickness amongst turkeys whatever in California, if they are properly cared for, and I think eventually California will supply the Eastern States with their Thanksgiving and Christmas dinners, for they have there a disease among turkeys which is so serious that it is decimating, and, in some places, wiping out whole flocks of turkeys. The disease is called "Blackhead," as the head in some instances turns black or dark colored before or at the time of death.

The Oregon Experiment Station has recently issued Bulletin No. 95, by E. F. Pernot, on Diseases of Turkeys. This bulletin contains information of very great importance to the turkey raisers of the state. It treats the subject of Blackhead, explaining the cause of this disease, the symptoms, and treatment. This bulletin, which may be obtained free on application to the Experiment Station, Corvallis, Oregon, should be in the hands of every turkey breeder in the state.

In sections of the East, Blackhead has almost wiped out the turkeys, and the same thing is liable to happen in this state if proper measures are not taken to prevent it.

I give here a brief summary of Prof. Pernot's bulletin:

Symptoms—Diarrhoea is the most pronounced symptom. The discharges are frequent, thin, watery, and generally of a yellowish color. This, however, sometimes occurs from other intestinal disorders, and does not alone signify the presence of the malady. The next symptom is the drooping tail, followed by a drooping of the wings, after which death soon ensues. When the disease is at its height, the head assumes a dark color, hence the name, Blackhead. Young turkeys are much more susceptible or they may be more delicate and cannot withstand the invasion of the parasites so well. They begin by moping and bunching up as though they were cold, diarrhoea soon sets in, the tails droop, then the wings droop, and they go about uttering a pitiful "peep," after which they soon die. A blackening of the head does not always occur.

It is only by careful post mortem that the true cause of the disease may be determined.

The Cause—The disease is caused by animal parasites, which can be detected only by the aid of a microscope. Because of their minuteness and growth in the mucous membranes of the digestive tract, they are easily carried by the excreta to food, which upon becoming contaminated, transmits them to other fowls. This is the usual means of infection.

Remedies—Food given to fowls should never come in contact with their droppings, as one bird with the disease will infect the feeding ground of others. Better sacrifice the bird at once than run the risk of spreading the infection to the whole flock. A sick bird should be removed from the flock and placed in close quarters, which may afterwards be disinfected, or the bird may be killed at once and then should be burned. Medical treatment is not very successful, owing to the difficulty of reaching the parasites at the seat of the disease; yet treating them with some of the following remedies is well worth the trouble: Sulphur, 5 grains; sulphate of iron 1 grain; sulphate of quinine, 1 grain. Place this amount in capsules and administer one night and morning to each turkey for a week. If the bird does not respond to treatment, kill it at once without drawing blood, and then burn the carcass, disinfecting the coop.

A solution of carbolic acid prepared by mixing five parts of the acid to 100 parts of water makes a good disinfecting solution, or chloride of lime, 5 ounces to 1 gallon of water, is good. Corrosive sublimate in the strength of 1 ounce to eight gallons of water, is a strong disinfectant, and may be used with a broom or spray to wet every part of the coop and floor, but it is poisonous and must be handled with great care. To disinfect the entire premises when the fowls are running at large is impracticable; but lime should be used freely on the droppings beneath where they roost. When the disease becomes seriously destructive, it is more than likely all the flock are affected, and it may be necessary to destroy all the remaining birds and disinfect the premises as thoroughly as possible. In such cases it would be better to suspend the raising of turkeys for one year.

Liver Complaint

Personally I have only met once with a case in California which might be called Blackhead. I have seen many cases of common liver complaint, and by my directions others have succeeded in curing many of these.

Dr. Salmon tells us that the seat of the disease called Blackhead is in the caeca. The caeca is sometimes called the blind bowel; it is a sort of "appendix" in the turkey, having no outlet. It is two lobes of bowel united by a ribbon of fat (the pancreas). In Blackhead and also in some cases of liver complaint, an abscess forms in one or both caeca, but this can only be discovered after death, and I have only found it in a post mortem of one turkey. The fact is, I have been so very "lucky" in raising turkeys that now I rarely even see a sick turkey, and I have many letters from our readers telling me they have cured their turkeys by my directions, so I will repeat them again for the benefit of newcomers.

First, liver complaint comes from wrong feeding, or overfeeding, which has overworked the liver; secondly, Blackhead comes from a parasite; thirdly, the symptoms of both diseases are almost exactly the same in the first stages. Dr. Cushman, in discussing this matter, decided that when the bright yellow diarrhoea comes on, showing liver trouble, the remedy is "something bitter and something sour." This is easy to remember. He also recommends no food but green food and says that turkeys have been known to cure themselves by living on acorns.

My remedy is first a liver pill followed by quinine for a week, and sour milk and no food but onions and green alfalfa or grass, keeping this up until cured.

I have a letter from a successful turkey raiser of Long Beach near Los Angeles. She writes: "I wish to tell you my experience with liver sick turkeys. I had a gobbler weighing eighteen or twenty pounds, and I made the mistake so many do of allowing turkeys and chickens to run together; my experience is that turkeys, especially toms, will not stand such quantities of food that hens do. Well, he got very sick, so bad he was as light as a feather, and my cure, which never fails—was administered—a bottle of Jamaica ginger and a bottle of liquozone were procured. I put him in a clean, large coop and he lay on a bed of straw for days, so weak he could not stand. The first day I gave him one teaspoonful of the ginger and one teaspoonful of the liquozone mixed and diluted until it was not too strong, giving two or three spoons every hour of the diluted. The next day giving it three times a day; after that twice a day. I did not allow him anything to eat, but of an evening gave him the smallest sized capsule of quinine. Kept that up until he

First, Second and Third Prize White Holland Turkeys, Mrs. W. D. Root, Glendale, Cal.

began to get good and hungry, then fed him a few grains of wheat, only about six grains, and a little speck of alfalfa. I have found that feed kills them every time when they are so sick. I never fail to cure the worst cases if I treat them like I tell you. Then if they hump up again and begin to get sick again, I give them a dose in the evening. The ginger warms them up and starts circulation, and the liquozone kills the germs."

Liquozone is very acid, it tastes like sulphuric acid and water, and I have no doubt that my friend's cure is a good one. Remember, Dr. Cushman says "something bitter and something sour," and if your turkeys get sick, try it immediately.

The Fattening of Turkeys

At this season many letters are coming to my desk either asking how to fatten turkeys or describing the ailments and often the death of the turkeys on which hopes had been based of a rich harvest of dollars for Thanksgiving and Christmas.

One writes: "I have followed the directions in your book with great success in raising turkeys, for I have not lost one, but some that I have cooped up to fatten won't eat and are not gaining in weight. Will you tell me how your fatten yours?"

I will willingly tell you, but first always remember my maxim, "When in doubt, consult Nature." How do the turkeys acquire the fat that they require to keep them warm during the winter?

All summer long and into the fall they have devoured grasshoppers and insects but with the chilly fall weather these are becoming scarce and the weed seeds are ripening, the nuts of all kinds are falling and the berries are at their best; Nature has provided with a liberal hand for the necessary winter fattening.

There is a flavor belonging to the meat of a range-fed or wild turkey that cannot be found in one raised in confinement, for necessarily the food cannot be so greatly varied, and the wild berries and nuts, the seeds of the pine cones, the beech nuts, hazel nuts, acorn, berries and spicy seeds as well as the buckwheat, barley, oats, wheat, corn, etc., impart a flavor not to be excelled, and the turkeys fed on these are fat enough for the most epicurean appetite. All that a free range turkey may need is a feed of corn at supper time.

For turkeys on limited range, or on range that may be bare of insects, nuts and berries, we may have to assist Nature and substitute for her fare the best thing that we can find and undoubtedly that is good hard corn a year old (so as to be thoroughly ripe and dry) for there is something in new corn which is apt to disagree with turkeys.

If at the same time the turkeys can be in an olive yard, where they can pick up the few olives that are now falling or may help themselves to some on the branches, this with a little corn at night will put them in fine market condition and is all the fattening they will need. The same will be the case if they can be in a sweet apple orchard, they are very fond of sweet apples, which agree with the turkeys and are also fattening. The walnut orchards would be good, too, but are usually picked so clean that there is nothing left for turkeys. The acorns in some places are as fattening to turkeys as to hogs and the wild nuts do not ever seem to disagree with the turkey's liver or digestion.

Turkeys that have not the advantage of freedom or the wild nuts and spicy berries can be successfully fattened in a yard. The way I feed is I take 2 parts of corn, and 1 part of barley, soak over night, and in the morning put on the stove to boil, let it cook slowly until it begins to soften, then take up and set aside covered till supper time, when it will be cool enough to feed. Give this three times a day as much as the turkeys will eat up in about fifteen minutes, then remove till next meal time. I add a chopped onion at supper time, as that is a stimulant to liver and digestion. Fresh water must be kept before them.

One year I had most excellent success in fattening turkeys by feeding them the same food as for fattening chickens, that is, equal parts of heavy bran, corn meal and oatmeal (rolled breakfast oats), mixed with milk or with buttermilk, three times a day.

Boiling the corn removes whatever it is in the new corn that disagrees with the turkeys, scalding the corn meal has the same effect. Adding a little ground charcoal will assist in the fattening and prevent indigestion. I have known it to materially increase the weight by enabling the turkeys to eat more or digest more food.

One thing in fattening turkeys, begin gradually by feeding the fattening food only once a day, and that preferably at the evening meal.

I only fatten turkeys from two or three weeks. The gain in weight depends upon the condition, size and age of the turkey, when commencing to feed. It will vary from one to, in rare cases, as much as six pounds. This last is claimed by feeders in Europe, where the cramming machine is used.

Do not fatten the turkeys you intend to use as breeders, for the fat weakens the organs of reproduction in both sexes and the offspring will be weak and small or the eggs infertile. This I have found almost invariably to be the case in many instances where people have had me investigate the cause of eggs not hatching.

One word about turkeys getting sick when being fed for market. It shows a lack in the constitution, and is often the result of a slight attack of liver trouble, which they seem to get over, in their early life. The best thing would be to turn them out on the range again, or to doctor them up by giving them a dose of Epsom salts and then following it with ten drops of tincture of Nux Vomica in a pint of water, allowing no other drinking water. Then give freely of chopped onion and bran mixed, as well as the fattening food, and eat them as quickly as possible. Do not breed from these turkeys, as, although they may be perfectly well, they will not breed a vigorous constitution into their offspring.

Turkeys should be kept at least twelve hours without food before killing. They may have water, but no food. This is to empty crop, gizzard and bowels, and prevent the food which would remain there from souring and giving the whole carcass a bad flavor.

DUCKS AND THEIR VARIETIES

In the springtime of the year in the East the big duck ranches hatch ducks by the hundreds of thousands, but in California, or at least in the neighborhood of Los Angeles, there are not such large ranches, and ducks do not seem as popular. Probably some farmers have had a few in their yard at some time, just to give them a trial, and have found them a continual nuisance, as they greedily eat the whole allowance of food from expectant chickens and dabble in their drinking vessels, so they have to be continually cleaned and replenished, and with great injustice to the ducks, they have let this prejudice them, where if they had kept the ducks separate, they would have found them easier to raise than chickens.

Ducks grow faster and are ready for the market earlier than chickens; they are not troubled by the diseases of hens, neither do

"WONDER" INDIAN RUNNER DUCKS. FROM 283 EGG STRAIN IMPORTED FROM NEW ZEALAND BY MRS. M. E. PLAW, FRUITVALE, CAL.

they have lice, except if raised under a hen when very young, before the feathers grow, the gray head-lice may get on their heads, crawl into their ears and kill them, but this is before they feather out. Mosquitoes, which are very troublesome in some places to the chickens, causing great mortality, never trouble ducks, neither do fleas or ticks. I think the reason for their immunity from vermin is that their feathers are very oily and thick and the down under the feathers is an extra protection. Hens require a dust bath, while ducks require a water bath to keep them clean and healthy.

Most of the popular varieties of ducks can be raised and bred without water to swim in, but on the very large duck ranches a supply of running water, so that they may have fresh water to drink and a bathing place for the breeding ducks, is a great advantage.

Ducks should be kept entirely away from chickens and turkeys, as they pollute water so badly it makes the other fowls sick. I found on my small ranch where there was only water piped in. after trying various plans for watering the ducks, an easy and convenient way. I had a barrel sawed in two, two-thirds and one-third. I knocked the head out of the larger end and buried that part, making it deep enough so the top of the barrel was just below the ground; any box with no bottom would do as well. The one-third of the barrel had a bunghole in the bottom. This one-third barrel I placed over the sunken one. I had a broom handle which fitted into the bunghole and every day I let the dirty water run through it into the bottomless barrel and it soaked away. In this manner I gave my ducks fresh water and a clean bath every day. I found if I sawed the barrel exactly in half, it made the top part deeper than I wanted, and the bottom not deep enough.

GOODACRE'S PRIZE WHITE INDIAN RUNNER DUCKS.

The Varieties

I have successfully bred the following most popular breeds of ducks and think a slight review of them may be interesting and helpful to beginners: The Aylesbury, Pekin, Indian Runner, Buff Orpington Duck and the Muscovy.

The Aylesbury

The Aylesbury, called after a town in Buckingham, England, are about a pound heavier than the Pekin. The standard weights being, drake, 9 lbs.; duck, 8 lbs.; young drake, 8 lbs.; young duck, 7 lbs. Their color is pure white, with pinkish-white beak and shanks. They are extremely popular in England and are hardy and vigorous. There are not many breeders of them in this country, but an Englishman, Mr. V. G. Huntley of Petaluma, who has imported some exceedingly fine Aylesbury ducks from England, says he has a large demand for them, as they are a rarity in this country. He considers their flesh better than that of any other variety of ducks. In plumage the Aylesbury are a pure spotless white, with hard,

AYLESBURY DRAKE.

close feathers that glisten in the sunlight like satin. The advantages claimed for this breed are the easiness with which it is acclimated, its early maturing, its great hardiness, its large size, being heavier than any except the Rouen, its great prolificacy and its beauty.

The Pekin

The Pekin is undoubtedly the most popular breed on the large duck ranches in the East, where thousands of them are fattened and turned off every season. This breed is variously called the Imperial Pekin and the Mammoth Pekin and Rankin's Pekin. It was brought to this country from China in the early seventies and immediately took the first place as the most prolific and rapidly growing

MAMMOTH PEKIN DRAKE. REICHART'S DUCK FARM, SAN FRANCISCO.

duck on the market. In shape and carriage the Pekin has a distinct type of its own, which by some is described as resembling an Indian canoe, from the keel-like shape and the turned-up tail. Though Pekin ducks may not merit all that is claimed for them by enthusiastic breeders, it is certain that the duck business could not have attained its present proportion without the Pekin duck, and that as a market duck this breed takes the lead. They are hardy, quick growers, thrive in close confinement, and are ready to market at ten weeks of age. The plumage is soft, more downy than that of other varieties and is of a creamy white in color. The beak is of a deep orange yellow, and, according to Standard, should be free from black marks. The shanks and toes are reddish orange color.

All ducks are of a timid disposition, and the Pekin more so than those of other breeds; in fact, they will injure themselves so badly if frightened by cat, dog or a stranger, or by being caught up, that they may have to be killed. A fright, if not fatal, will take off several days' growth of the young, and stop the laying of the adult ducks.

The Indian Runner

Many years ago Indian Runners were brought from India to England by a sea captain, hence the name "Indian," while the "Runners" came from their great agility. They do not waddle like other ducks, but run more like a plover, and are very quick in their movements. In England their good qualities quickly captivated the thrifty farmers. Individual ducks there have made a record of 225 eggs per annum. Here in California I had ten ducks which laid 2331 eggs in one year. I think the climate of California more nearly resembles that of their native land, and their laying is never checked by cold or snow, so that here they lay better than in England or the Eastern States. In India they were bred for their laying and table qualities, no attention being paid to the color of their plumage; all the Indians cared for was the eggs, and they laid eggs galore. English breeders claim that eight-year-old ducks of this breed will lay as well as yearlings, and on this account, and their capacity for foraging, they have become very popular in England and Australia.

While the weight of the matured Pekin is greater than that of the Indian Runner, there is more meat in proportion to their weight in the Runners on account of the smallness of the bones; the meat is also of a much finer quality, finely grained and juicy and resembling in flavor the much extolled canvas-back duck. The eggs of the Indian Runner are an ivory white in color, greatly resembling Minorca eggs, very delicate in taste, and in England their eggs are in great demand in the tuberculosis sanitariums on account of their delicate flavor, richness and nutritive value, and absolute freedom from tuberculosis taint, and there is a higher price paid for them than the hen's eggs.

The standard color of the Indian Runners in this country is fawn and white and pure white. In England they also have the black and white, the brown and white and the pure white.

The pure white ducks are meeting with great favor in this country and are becoming very popular and are said to be as good layers as the fawn and white.

The Rouen

The Rouen duck, so named for a city in Normandy, where they are supposed to have originated, are still bred there in large numbers. The Rouen duck is a fine market bird, but does not mature as early as the Pekin or Aylesbury. It is easily fattened, hardy and quiet in disposition and not as nervous as the Pekin.

The Rouen drake is a magnificently colored bird. Neck and head are iridescent green, breast wine color and the lower part of the body delicate steel gray, penciled with very fine black lines. About June a remarkable change takes place in the drake. He begins to lose his lustrous feathers, those of the neck dropping out, being replaced by feathers of a russet brown. The magnificently colored drake is clothed in sober hues for the summer. In October he again resumes his gorgeous raiment.

The Buff Orpington

Buff Orpington ducks are a breed of Mr. William Cook's making. He named them as he did the Orpington hens, after his own place in Kent, England. The color of the Buff Orpingtons is a soft shade of buff, the drakes having rich brown heads. The Buff Orpington has a good deal of the Indian Runner blood in it, and from this source its laying qualities are gathered. Mr. Cook claims they are better layers than any other of the duck family. Many of them lay a beautiful green egg, although a greenish-white is the usual color. These ducks weigh a pound a half more than the Indian Runner, are large and more plump birds, maturing early, and one of the best market birds.

The Muscovy

The Muscovy duck is not largely bred in this country. They are not like any other ducks and do not interbreed with others. It is a native of South America, where it may still be found in its wild state. It comes in two varieties, white and black and white. The males are much larger than the females. I had one weighing fourteen pounds. Both sexes have caruncles at the base of the beak; these become larger every year, giving them a vulture-like appearance. Muscovy ducks are rather awkward in the water, preferring to live on the land. They are pugnacious and ill-tempered, and, although they have web feet, they have very sharp claws that can, and do, scratch in a most unpleasant way. They are strong on the wing, flying easily over the barn, and they like to perch on the roof. They are good setters, and their eggs take thirty-five days to incubate.

Hatching and Brooding

The first thing the amateur needs is first-class breeding stock or eggs of the same. There is sure to be sad loss among young ducklings, bred from debilitated stock. Good stock should be secured to start with, and when properly fed and cared for, there need be no fear of loss.

A good incubator, carefully operated without variation of temperature, should receive the eggs. They take twenty-eight days to hatch. Duck eggs will hatch well in any of the standard incubators; they require more airing than do the eggs of the hen, and I have found that by sprinkling them every other day, after the first week, I was sure of a good hatch. Sprinkle the eggs, or moisten them thoroughly, with warm water, when they are out of the machine, and do not put the water in the incubator. I found this much the best plan. I think wetting the shell of the egg helps to soften it and make it more brittle, enabling the duck to break its way out easily. I also do this when hatching duck eggs under hens.

A brooder adapted to chicks will answer equally well for ducks. The little fellows should be at least thirty-six hours old before taken from the incubator and placed in the brooder, which should be previously prepared for them by placing a board about ten inches wide a few inches from the front of the brooder forming a very small yard with a little water fountain so arranged that they can get their bills in but not their bodies. The birds should be confined to this small space in front of the brooder for the first day, or until they have learned the way into the hover. Bed the little fellows with hay, chaff or cut straw. Keep the pens clean, both outside and in. The welfare of the ducklings depends upon this. Be sure to give them shade.

Mr. James Rankin has been called the father of the duck industry in America. He and a number of others in the East are now hatching by the thousands and tens of thousands. He writes: "With us it is the surest crop we can grow; it makes the best returns of any crop on the farm."

As he is a noted expert in the business I cannot do better than give his directions for raising the ducks and his formulas for feeding at the different ages. I have tried them myself and do not think they can be improved upon.

Feeding

The first food should consist of bread or cracker crumbs slightly moistened and about 10 per cent of hard-boiled eggs chopped fine, shell and all; mix in this food five per cent of coarse sand. Do not place grit by them and expect them to eat it, but mix the sand in their food and so compel them to eat it as it is the most essential part of the whole thing.

Scatter the food on a board, place the young ducklings on it and they will be busily eating it within ten minutes. One hundred to one hundred and fifty ducks can be put in one brooder six feet long. When two or three weeks old, not more than seventy-five should be kept in one brooder. The heat under the hover should be kept at about 90 degrees for the first day or two, when it should be gradually reduced as the ducks grow older. In the climate of Southern California, ducklings rarely require brooder heat more than two weeks.

The second day rolled oats and bran can be added to the food; a little finely cut clover, lettuce or cabbage can now be safely used. At ten days feed one-fourth corn meal, the rest wheat bran with a little rolled oats mixed in, not forgetting the grit, about ten per cent of ground beef scraps, and the same of green food. At six weeks Quaker oats, grit and ten per cent beef scraps; at eight weeks old feed equal parts of bran and corn meal with a little Quaker oats, grit and beef scraps, but no green food.

The birds should be ready for the market at ten weeks old. They should be fed four times a day until six weeks old, then three times is sufficient. They should be watered only when fed until six weeks old, then they should be watered between meals also. Feed at each meal all they will eat up clean, then take the remainder away; keep the pens dry and clean and be sure you give them shade.

For breeding birds, old and young, during the summer and fall, when they are not laying—feed three parts wheat bran, one part Quaker oats feed, one part corn meal, five per cent beef scraps ground fine, and five per cent coarse sand, and all the green feed they will eat in the shape of corn fodder cut fine, clover, or oat fodder, or alfalfa. Feed this mixture twice a day, all they will eat.

For laying birds—equal parts of wheat bran and corn meal, twenty per cent of Quaker oat feed, ten per cent of boiled turnips or potatoes, fifteen per cent of clover rowen, alfalfa, green rye or refuse cabbage chopped fine and five per cent of grit. Feed twice a day all they will eat, with a lunch of corn and oats at noon; keep grit and crushed oyster shells before them all the time.

Mr. Rankin adds: "I wish to emphasize several points. Do not forget the grit, it is absolutely essential. Never feed more than a little bird will eat up clean. Keep them a little hungry. See that the pens and yards are sweet and clean, for though ducklings may stand more neglect than chicks, remember that they will not thrive in filth. If anyone fails in the duck business, it must be through his own incompetency and neglect."

Mr. Rankin has his yards swept twice a week. These sweepings amount to many tons each season, and are spread evenly over his grass farm, giving enormous crops of good hay, so that where, twenty years ago, only six tons of hay were cut, now the crop is 125 tons.

On Long Island the method of feeding is as follows:

From the time of hatching until seven days old, feed equal parts by measure of corn meal, wheat bran and No. 2 grade flour. This grade of flour is sometimes called "red dog" flour. To this add 10 per cent of the bulk of coarse sand. Mix with water to a crumbly mass and feed four times a day.

From seven to 56 days feed equal parts by measure of corn meal, wheat bran and No. 2 flour; 10 per cent of this bulk of beef scrap; 10 per cent of coarse sand and about 12 per cent of green stuff. Mix and feed as before. From 56 to 70 days feed 2 parts by measure of corn meal, 1 part wheat bran, 1 part No. 2 grade flour; 10 per cent of this bulk beef scrap; 12 per cent green stuff. Mix and feed as before.

It should be remembered that both green stuff and beef scrap are absolutely necessary to the best growth of ducklings, and no one should undertake to raise them without feeding both, as ducks deprived of them never make as good growth as those which are supplied with them. Mix the feed fresh for every day in a trough, and if the weather is hot, mix twice a day. Keep the mixing troughs clean and sweet. Feed in troughs, giving at each feed as much as will be eaten clean before the ducks stop eating, and no more. A little observation will show how much to feed.

Ducks that are to be reserved for breeders should not be forced as rapidly as those to be sold in market. While the rations for breeding ducks should be rich in protein, they should not be such as to produce a surplus of fat. In raising breeding ducks the object is to secure large size, which needs a large frame, thick muscles and great vitality. For this reason less corn meal is fed. An excellent ration for ducks reserved for breeders:

Equal parts by measure of corn meal, wheat bran and green stuff, with 5 per cent of beef scrap and 5 per cent coarse sand or grit.

Ducks are good grasshopper catchers and industrious insect hunters, but they should be given beef scrap regularly, even when they have their liberty.

Ducks are profitable on the farm, as they are good layers, make weight economically and are always in demand in every market. Such a thing as overstocking the market with ducks has not yet occurred, and many farmers might keep a small breeding flock and raise 100 or more ducks every year to advantage.

We cannot close the chapter on duck feeding without drawing attention to several important points.

The first is, that with ducks especially, all food should be given on boards or troughs, at any rate not thrown on the ground, or it will become very foul. When the trough or board is not in use it should be stood on end alongside of the fence, otherwise the birds will get it dirty.

Fresh water must be provided for ducks, deep enough to immerse their nostrils, and the vessel must be large enough so that the supply will not run out. Ducks must have water always before them; to go without even for a few hours is worse to them than missing a meal. One will never make a success with ducks without provision for a constant supply of water.

Another matter that must be attended to is the supply of grit or coarse sand and crushed oyster or clam shells. One has only to experience the keen stoppage of eggs that follows the running out of the grit supply to realize its vital importance.

A feeding trough, a good sized water vessel, and a box of grit are about all the furniture necessary for the duck pen.

Breeding Ducks

In mating your breeding pens, special attention should be paid to the male. He is in theory and in fact half the pen. Every young duck reared will be half his blood and will to a great extent take after him. Where there is a faulty female, only her own progeny will be affected, with the male it is entirely different, consequently the male should be the best in health and vitality that you can get. You can afford to pay a good price for him if he makes every duckling worth only a few cents more than the ordinary ducks. You will find that it will pay to use young drakes (from eight to ten months of age), whilst the age of the ducks (especially Indian Runners) does not so much matter. The fertility is always better with a young male, especially early in the season.

One great cause of infertility is overfatness of the breeding stock. A bird to lay well must be in good condition but not overfat. There may be said to be two kinds of fatness, one we might call soft and the other hard fat. When a duck has been fed a too carbonaceous ration it is either passed away as waste or is stored up in the body as hard yellow fat, which may largely interfere with the sexual organs which become displaced or obstructed, with the result of infertile eggs. The birds are too fat, that is they have had too much fattening food whilst they may be almost starving for nitrogenous food, which will if used in conjunction with the fat be manufactured into eggs. Improper feeding, not overfeeding, which is practically impossible if the food is of the right quality, or as we call it "properly balanced," means not only a loss of eggs but a loss of fertility.

When the ration is too fattening it will often be noticed that the eggs are misshapen or are too small or too large.

Lack of exercise is also a cause of infertility, and the best way of rectifying this is either to give the ducks a good grass run, or a pond in which to disport themselves for at least a portion of the day.

The proper number of ducks to be mated to one drake varies according to the season and the breed. From three to six ducks for the Pekins and from eight to ten or twelve Indian Runners has been found the best number. Several drakes can be kept in the same flock, as they do not quarrel and fight as do the chickens. It pays best to sell off the males at the end of the breeding season, except in the case of special show specimens.

In hatching duck eggs we have to rely either upon a good incubator or upon hens or Muscovy ducks, for the domesticated duck does not go broody, and the rare specimens that do want to sit cannot be relied upon. A good incubator, operated without variation of temperature, is most generally used in this country. Duck eggs take twenty-eight days to hatch. They require more airing and cooling than the chicken eggs, and I have found it best to sprinkle the eggs or moisten them thoroughly with warm water and not to put water into the machine; I also do this when hatching with hens. By this plan I did not have ducks drowned in the shell which is usually due to having the water in the incubator and not airing the eggs enough.

The proper airing depends greatly upon the weather, so no set rule can be given, but I generally aired them the first week, after the first four days, for ten minutes; the next week for fifteen minutes a day, and after that for twenty minutes, whilst the last week up to the twenty-fifth day I aired them for a full half hour. It depends upon the heat of the weather. I have had the eggs left out accidentally for three or four hours and had a good hatch. I think that the principal cause of poor hatches is improper care and feeding of the breeders. Breeding ducks should have an abundance of green food daily.

Muscovy ducks are most excellent incubators. They are used as incubators both in France and especially in Australia. In these and possibly in other countries they hatch turkey eggs, duck eggs and even chicken eggs. In some places in Australia five hundred Muscovys are kept for sitting on duck eggs, as it has been found that they hatch out a much larger per cent of eggs and with comparatively little trouble to their owners than either hens or incubators.

Muscovy duck eggs take thirty-five days to hatch, consequently they make very patient and steady sitters on eggs and will hatch duck, turkey or goose eggs without difficulty. In using Muscovys you will probably need one Muscovy duck on an average to every thirty youngsters you wish to raise. Actually, they will hatch and raise a great many more, but it is as well to give a low estimate. The Muscovys on this coast only need an open shed with straw; you can keep the flock together. They will not interfere, but each female will build her own nest. They make their nests on the ground by hollowing out a hole with their bodies and lining it with straw. When the ducks are about to sit, they pull feathers from their own breast and with these line the top of the nest, so that one may always know when a Muscovy duck is ready to sit. A Muscovy duck will cover from twenty to twenty-five duck eggs and will brood from forty to fifty little ducklings. When the Muscovy duck leaves her nest to eat, which she will once or twice a day, she covers up the eggs with the feathers and down. Towards the end of the hatch she will often stay off the nest a full hour without injury to the eggs.

Muscovy ducks make excellent mothers, or you may say brooders for turkeys, ducks or chickens, on account of their large wings and very warm bodies.

SOMETHING ABOUT GEESE

Geese are, of all fowls, easiest to raise where grass is abundant, for they are grazing animals. Among the various breeds raised in this country the Toulouse is the most profitable goose to raise. It grows the largest, matures the quickest and is not so much of a rambler or flyer as the other varieties, and as it does not take so readily to water it grows more rapidly and accumulates flesh faster than other varieties, and is not so noisy.

There seems to be a steady demand for the beautiful large, gray Toulouse variety. They deserve every word of praise given them. They have been known to live to a great old age. I have had a friend in England who had a goose that had been more than a hundred years in the same family, and even at that age produced as many fertile eggs as any in the flock. In fact, that goose had more broods each year than any other goose in the neighborhood.

There are many points about raising geese that can be learned only by experience and a little practice is worth a world of theory. Intelligent and systematic breeding is sure to bring both pleasure and profit to the breeder.

Hatching and Feeding

For hatching goose eggs, if setting hens are used, keep them free from lice by dusting with insect powder every week, and put from four to six goose eggs under every hen. After eight days test-out, leaving four fertile eggs under every hen to hatch. Goose eggs should be sprinkled every fourth day after the twelfth, with warm water. In hot, dry weather, float them in water for one and a half to two and a half minutes. If incubators are used, float always. At the last float hold the pip up so as not to drown the gosling inside the egg. If the gosling remains and dries in the shell, it should be helped out. Break away a little of the shell, and if the lining does not bleed the gosling is ready to come out. Wring out a cloth in water as hot as you can bear your hands in, wrap the egg in the cloth and leave for a few minutes. You will find the gosling will come out bright and clean. Keep the goslings warm until they are dry and can run around. When they are twenty-four hours old put them in a box, the bottom covered with sand, and feed them often with a crumbly mash of one-third corn meal, two-thirds bran and a pinch of sand.

Goslings are Healthy

No other young in the whole tribe of domestic poultry is so up-to-date and healthy as a young gosling. Given a tender grass plot and a bit of warmth, it goes merrily on its way, nipping a living and asking favors of no one. They eat daintily, preferring grass to all other foods. With their chatter they are ready to meet you, take a few mouthfuls of food, and, with the same old tune, they lazily saunter away in search of grass and more rest.

Geese are turned out to pasture just the same as cattle, their bills having serrated edges which enable them to graze. They never need a warm house. An open corral is much better in California for them and they are not given to disease. Goslings, however, should be provided with shade, as they suffer from heat, getting a species of blind-staggers or sunstroke if exposed to the sun.

One of the best items of profit to be derived from a flock of Toulouse geese is the feathers, which are clear gain, costing nothing but the trouble to pick them. Watch them in the fall and spring, twice a year, when they begin to pull out the feathers and throw them away. I know then they are ready to pick. I think it is cruel to pick at any other time. Make cheesecloth sacks which will hold two pounds of feathers. Make them large, as the feathers will cure better if they are not packed together. Hang the sacks on a clothes-line every sunny day for about two weeks, then keep them in a well-aired room. Women living in the city will be your best customers, providing you let them know you have good feathers for sale. One can get from 75 cents to $1.00 per pound, and can never supply the demand. The breeders should not be picked when they are laying.

The Varieties

There are a number of varieties of geese, but the most profitable are the Toulouse, the Embden, and the China. Of the latter there are two kinds, the brown and the white. The color of the Toulouse is gray and white and the Embden is white. The Toulouse and the Embden are the larger. A pair of Toulouse have been known to weigh 59½ pounds, and an Embden pair has tipped the beam at 57 pounds. They are great layers of large eggs, of which they will lay thirty to forty a year, although I know a woman who has a goose that laid 70 eggs without wanting to sit.

In mating, allow two geese to one gander, though they generally pair off and the gander will stay with his actual mate nearly all the time. The gander is the protector of the goose, especially in breeding time. He will defend her and her nest fearlessly.

Hens as Mothers

It is a good plan to put goose eggs under a hen. It takes thirty-one days to hatch them. Then you want to be on the watch. The hen will sit all right, but when the young ones break the shell and the hen sees a queer, green little creature, with a long, wide bill saluting her, she takes it for a freak of nature, and off comes its head. Not many hens will claim the young geese or hover them; so take the goslings away as they hatch and try the hens, giving the goslings to a good, slow, gentle hen. As soon as she takes them without any fuss there is no danger. If the weather is nice they should be turned out in a small enclosure, which can be changed every day or so. Use boards six feet long and twelve inches wide. After a week let them go, and their foster mother's trouble begins. The little goslings do not care for her calling; they are hustling for every spear of grass and she has to hunt them. Her business is to keep them warm at night and warm them in the daytime if they get chilled. Never allow goslings to get to water to swim until they are fully feathered, and then only let those go that you wish to keep for breeders. Many of them will do as well if they never go swimming. During this period you must keep the old geese away, as they will fight the hen and molest the young.

You cannot raise geese as you do chickens and ducks, on a city lot. They must have pasture. It is a wrong belief that geese or their droppings will kill grass or pasture. If you have a large flock of geese and a small pasture they will clean it up; that is, they will eat the grass as fast as it sprouts and give it no chance to grow, just as a cow on a city lot will soon have only bare ground and you will have to tie her in the road. If you do the same with geese you would find the grass growing again the same as before. Geese are easier to raise than any other young fowls.

Muscovy ducks make excellent sitters for goose eggs.

.CAT AND HAWK-PROOF COOP FOR CHICKS AND DUCKLINGS.

PHEASANTS.

It takes time, patience and energy to raise pheasants successfully. Any successful poultry raiser can succeed with them, although they are not as easy to raise as chickens, but by following as closely as possible to Nature's way we can have good success. Pheasants are hardy, strong, very prolific and when young are quite tame. Pheasants do not hatch their own eggs in captivity; when wild they make excellent mothers, but captivity destroys the hatching instinct except to a very limited degree.

The price of the pheasants has a good deal to do with the choice of a breed. The Chinese, English and Golden are the most in demand. These are the heaviest egg producers. The Silver, Reeves, Amhersts and Swinhoes are close followers. The Chinese pheasant is usually the cheapest. The Golden is somewhat smaller than the Chinese, is tamer and more brilliantly marked. The English pheasant is very similar to the Chinese, but rather larger, has a less conspicuous white collar and lays a larger egg. The English and Chinese bring about the same price. The other pheasants being less hardy and having a far less egg production bring higher prices.

The building of the pens for pheasants should be carefully done. One-inch mesh should be used at the bottom of the fence for two or three feet up. Although some people prefer boards two feet high at the ground, this is a good idea, as it prevents fighting between the different pens, also it keeps the young pheasants from wandering.

The pheasant pens should be located in as dry a location as possible, for the birds love their sun and dust baths. There should also be trees or bushes in the pens, where they can shelter from the sun and also hide away from people. A brush heap is their delight, and they will hide their nest in it in preference to elsewhere. The eggs should be gathered twice a day. They lay about twenty eggs at a clutch and then rest a little, in captivity.

The pens should be covered over with two-inch chicken wire, as pheasants fly like wild birds, although where they are kept for pets only the outer long pinion or flight feathers of one wing may be cut to prevent an extended flight. Care must be used not to cut the inner feathers of the wing, as these protect the bird's lungs. Pheasants are great runners and enjoy running about and slinking through the brush of their pens. The English and Chinese pheasants are polygamous, the same as chickens, and the male will even mate with wild grouse or with barn fowls.

The natural food of the young pheasant is insects of all kinds, larvae of grubs, worms and especially ants' eggs, as well as small seeds. In raising the little ones, use the same food as for the little turkeys at first, or, in other words, imitate the food that Nature provides for them. Be sure to give them chopped-up lettuce and onion, and a little later on the chick feed, but with a very small amount of corn in it, for corn does not agree with them. Corn is not their natural food.

After pheasants are three months old they are very hardy, and at five months are in their full plumage. The proper food for grown birds is wheat, heavy oats, buckwheat, clover, alfalfa and grass. They also dearly love raw apple, potato, cabbage, carrot and lettuce. Their preference, however, is flies, grubs, bugs and worms. They need plenty of good, clean, fresh water.

Anyone wishing to go into the pheasant business should write to the director of documents, Agricultural Dept., Washington, D. C., and get the Bulletin on the "Raising of Pheasants in the United States," also the book on pheasants by "Dillaway" at the Dillaway Pheasantries, Everett, Washington.

GUINEA FOWLS

Guinea fowls are becoming popular in this country and will be more so every year, as their excellent table qualities are more known.

Guineas are used to replace pheasants at banquets and at the closed season. They are sometimes passed off as grouse or pheasants, although at some of the large restaurants they are often given their own name on the bill of fare.

They lay a small egg, brown in color, with dots or little spots of darker brown, and quite pointed at one end. The eggs are considered a great delicacy in Europe, for they are very rich in the color of the yolk. The guinea hen lays a great number of eggs, but she is wild and hides her nest and two or three eggs should be left in the nest as nest eggs. I have kept guineas nearly all my life, and after being well acquainted with their habits I never touch the eggs in the nest with my hand, as they so dislike the smell of a human hand that they will desert the nest and it is often a trouble to find them. I always use an iron spoon to collect the eggs.

The male and female guinea are identical in color and can only be distinguished by the wattles of the male being a little larger and the "song" different. The female has a harsh voice, which calls "come back," "come back," whilst the male only seems to say "quit," "quit." This is when they both are comfortable and happy, but let a hawk appear on the scene and the scream of anger, definance or warning will cause every chicken, turkey or guinea on the place to run to shelter. Guineas are as good as a watch dog, night or day they will give notice if a stranger comes on the place. I have had male guineas that would fly into the air to meet a hawk and give fight.

Guineas can be hatched under common hens, and, indeed, that is the best way to start with them, as they are very "conservative" in their habits. They can be raised exactly as chickens, with one exception: they need food as soon as ever they are hatched. The eggs take 28 days to incubate and the little ones are exceedingly wild and will run away and get lost as soon as they are hatched if not closely watched. They should be confined in a tight pen, with sides at least fifteen inches high until they have learnt to follow the mother hen, which will be in a few days. The guineas soon learn to love their mother and will never leave her, in fact they will stay with her and roost with her even after they are laying eggs and are a year old. They are very peculiar in another thing, what one guinea does they will all do. If one flies over the fence, all will follow, a short of follow my leader game is going on all the time. The mother hen is followed by the young even after she begins to lay they will all go on the nest with her, no matter how she may peck them. I have had them effectually break up a sitting hen. They will often, if brought up with or by hens, lay in the same nest with the mother hen, although if at liberty, as on a farm, she will usually hide her nest. Guineas are gradually becoming polygamous, and the male will take as many as three or even four wives. The female makes a poor sitter and not a good mother. The wild nature is the cause of this, and if left to hatch her eggs and raise the young, she rarely brings more than four or five to maturity, at least this has been my experience. In the West, guineas begin to lay about April and continue until August. They weigh about three to three and a half pounds, and there is a growing market for them.

J. BURROWS' MODEL BREEDING ROOM FOR CANARIES, OCEAN PARK HEIGHTS, CAL.

CANARIES

Canaries can scarcely be called "Poultry," but all my life I have been a successful raiser of them and I so dearly love them that I want to give them a chapter in my book to let others know about the prettiest and dearest of "our little feathered brothers of the air." The wonder to me is that so few of them are kept and loved in California.

In England, in France, in Germany, and in most of the European countries, canaries are a source of income as well as of pleasure to the artisan or mechanic class. I have known personally shoemakers, plumbers, harness makers, carpenters, who were really expert canary breeders. I have myself patiently turned a little bird organ for hours day after day to reach the youngsters to whistle a certain tune, and at one time I took a number to the forest of Montmorency so that they could learn the nightingale song from the wild nightingales of the forest.

Canaries were brought to England about three hundred and fifty years ago from the Canary Islands. Since that time they have been extensively bred as household pets. During the three hundred and fifty years of its domestication the canary has been the subject of careful artificial selection, the result being the production of a bird differing widely in color of plumage and even in size and in form from the original wild bird.

In England, as well as in other countries, canary breeding is a hobby. There are hundreds of canary shows in England, and thousands of the lovely little pets are exhibited annually, but the climax show is at the Crystal Palace every February, when the champions from all over the country meet, and the judges have a hard time to select among so many almost perfect specimens the best canary in all England. The winner of each class in that Crystal Palace show means the choicest of ten to twenty thousand from over all England, Scotland, Wales, and even Ireland. The enthusiasm in England over these beautiful little peets is greater than in any other country, and it is not surprising, when one realizes that the expense of feeding a canary is next to nothing, and the care of them only a pleasure.

The price of canaries in England varies from one dollar to five hundred, for it all depends upon the beauty and the singing quality. In some places there are singing contests for some varieties. For show and singing and for general excellency the Norwich Canary carries the palm. It is the favorite breed in England.

I have asked a canary expert breeder and judge to tell us more about the different breeds of canaries that are popular in England and in Europe.

The judge (John Burrows) was for a number of years secretary of the Leicestershire (England) Ornithological Society and of the celebrated Leicester Shows, and knows more about the English canaries and English wild and cage birds than any one I have met in California. His description of high-class canaries will interest many.

The largest canary is the Lancashire "Coppy." This we place first because it is the giant of the canary family, often measuring seven and a half to eight inches long. As their name implies, they are bred for their crests, "coppy" being the old English for crest or cap. A good exhibition bird should have a drooping crest with a well-defined center, the crest feathers entirely covering the eyes and beaks.

Yorkshires—These birds are sometimes called the "Aristocrats" of canarydom. They are very straight, long and slim, with an erect carriage and feathers like wax, lying tightly to their slim bodies.

Lizards—A bantam variety of the canaries with beautiful spangled coats and a clear (not spangled) cap, the yellows are called the Gold Liazrds and the Siliver Spangled have the tips of the yellow feathers just slightly tipped with white. They are both most charming little, tiny birds.

Borders—These are another of the Wee Gems. They should be round in shape and as tight in feather for exhibition as though carved from boxwood.

Cinnamons—A beautiful variety of the Norwich type is called Cinnamon. They should be of sound cinnamon color with dark penciling showing on their coats. This is a truly grand variety.

Scotch Fancy—The Scotch Fancy is a great favorite in Scotland. They should stand with the head over the perch and the tail under, forming a crescent like a new moon.

Belgium—This variety should stand on the perch with the tail in a perfectly straight line with the back, head bent down so that the shoulders are the highest point on view. They are trained to keep the head down and the shoulders up, so that they have almost the appearance of being hump-backed.

DUTCH FRILLS—These are frilled on the chest, which makes them appear rough, but they have many admirers.

LONDON FANCY — This once popular bird, with clear, bright yellow body and dark wings and tail, seems to be getting quite scarce, now very few being bred, probably on account of the difficulty of breeding with the proper markings.

NORWICH—Last, but not least, we have the Norwich variety. This is without doubt the most popular and beautiful of the canary family, not only on account of the lovely and wonderful coloring, but also as singers they are second to none. In no other variety is there such depth of color, and so large a variety of markings.

The Norwich canary should have a full, round head with thick, short neck set on a chubby, round body, broad chest, short wings and tail. He should stand well across the perch with a bold, jaunty appearance. They vary in coloring and feather through all the various phases of marking from the green to the clear yellow, sometimes a specimen is seen with both eyes and wings marked alike, these are called "even marked" and are highly valued, while rarer still is the bird marked on eyes, wings and each side of the tail, this is called a "six-pointed" bird.

The Norwich is divided, as are the other varieties, in to "yellows" and "buffs." Good exhibition specimens of yellows are of the brightest orange, almost the color of red-hot iron. The "buffs," although as deep and bright in color, have the end of each little feather just tipped with white very lightly. The deep bright yellow shining through gives the appearance of being frosted, or as if a thin white lace veil were over the bird with the yellow shining through. The effect is almost indescribably beautiful. Some of these birds are extremely valuable. They are good songsters, exquisite little birds, bright and intelligent.

There are many shows in the different cities and towns all over England, Scotland, Wales and Ireland in the fall and winter and when these shows are over the owners of the prize winners send the champions (those that have never been beaten) up to the biggest and best show of all in the month of February at the Crystal Palace. At the 52nd annual show last February there were in all 2600 cage birds competing for prices. A prize there means a winning over about 26,000 birds from all over the country.

In reply to the request of one of his customers for a few instructions in the management of canries, Mr. John Burrows writes:

I like a room with the window in the East or Southeast, so that the birds will get the sun's rays when it is not too warm in the morning, also we must remember that the birds feed their young at the earliest opportunity in the morning.

I prefer a breeding cage as plainly made as possible, with just a wire front, every crevice must be puttied up, so there is no room for red mites, no ornaments or mouldings of any sort, it should be either limewashed or painted, with a drawer at bottom to clean them out, sliding divisions in the middle. Some fanciers use a slide with a few wires, so that the birds can get acquainted that way. Plenty of sharp grit in bottom of cage, the seed hopper in the center over slide, drinking water at each end, never put drinking water inside; four perches, two each side the slide, perches made big enough for birds to grasp firmly (no pencils). Hang nest box between perches when birds are ready to build. Size of breeding cage outside measure 36 in. long, 18 in. high, 12 in. deep. When your cages are ready, place the female on one side and the male on the other, when the male bird is seen feeding the hen through the wires, they are then ready to begin housekeeping, but don't put them together till March 1st, then you have plenty of time for three nests before the end of July. The male bird I sent you is a buff marked, he should be paired to a yellow hen.

Don't feed anything but plain canary seed in the hoppers, just a pinch of the following every other day in rotation, maw, rape, flax, hemp, millet; never give mixed seed, just a little tender green food every alternate day.

You will tell when the hen is about ready to lay, as she will finish her nest, and will not let the male pull it to pieces as perhaps he has been doing. When she has laid, take out the egg and put in a dummy, a small marble. She will perhaps lay every morning till the nest is complete, but if she misses a day don't worry. On the evening of the 3rd egg put them all back in the nest. She ought to start then and sit, they should all hatch together on the 14th morn. If the male bird is quiet he need not be removed, but if he is too gay he had better be put the other side of the slide until the young are one week old, when he can be put back and take his share in feeding and raising the family. As soon as the young leave the nest, the hen will be ready to lay again. By the time the second ones are one week old, the male bird can be returned again, as the first young ones will be able to feed themselves, and must be turned into a long flight cage, where they will have plenty of room for exercise. All the while the birds have young they should be fed on egg and bread food or ground crackers; nearly all egg when birds are first hatched, then increase the crackers as the young get older, also plenty of green food every day. Don't take their other seed away. They are also very fond of wild seeds, partly ripe, which is one of nature's foods for young birds. Soaked seed is also good, but it soon sours in this country. Everything must be kept as clean as possible, scalding all nesting materials, and be always on the lookout for the red mite. It is a good plan to dust the nest, when she begins to sit, with insect powder, also a day before she hatches.

I hope by following instructions in these few lines you will be able to raise quite a family.

GERMAN CANARIES—At my request Mr. J. C. Edwards, manager of "Birdland," Los Angeles, California, kindly sends me the following:

The people who have devoted greatest attention to the rearing of canaries are the Germans. By them the cultivation of the singing qualities of the bird has been almost the exclusive desideratum, no particular reference being made to beauty of plumage, shape or size. The finest singers in the world are the trained German birds.

The length of the Germany canary is about five and one-half inches, the color varying from pure yellow to a yellowish green. The birds are sometimes mottled or crested, for, as before stated, their breeding has entire reference to their song and not to their plumage.

In many districts of Germany the breeding of canaries is the principal occupation of the people, but the "Hartz Mountain" region surpasses all others in this business. The choicest breed is reared on the summit of the mountain in the little hamlet of "St. Andreasberg," where the bird education is carried to a degree that can scarcely be understood by the general public in this country. Every facility is given for the young birds to acquire the cultured notes of well-selected singers. Various mechanical devices are employed to introduce long trills, and flute notes and other oddities in song. From three to six months of constant training is needed to bring the young songsters to perfection. St. Andreasberg Rollers is the name of these canaries.

German canaries being bred by thousands of small breeders all over Germany, no one can tell just how many are produced annually. Our eastern importers dispose of over three hundred thousand of these birds in this country every year. Personally, I cannot see why we do not produce our own birds. We excel all foreign countries in superior poultry, horses, flowers, and by giving the matter the attention required we should be able to breed in time as good, if not better, songsters than any of the imported. There are thousands of people in this country raising poultry that they are selling for $1 a head and think they are doing well. Canaries can be produced for less than chickens, and will always bring more than a dollar. There is always a demand for good songsters and any one that will take up this matter on a sensible commercial basis can do well with it.

The rearing of young ibrds is a task in which all will not be equally successful, but it is safe to say that by following a few simple directions success will be assured. The breeding of canaries may be commenced about the middle of February and continued till midsummer, one pair raising several broods if permitted. However, continuing the breeding season too long is not advisable, as it will prove detrimental to the health of the birds.

The cage in which the breeding takes place should be roomy, sixteen inches in length is the smallest and ten inches in width, but larger cages result in better, healthier birds. The cage should be hung against the wall or placed upon a shelf some seven or eight feet from the floor and facing the south. When once it has become the home of the pairing birds its position should not be changed, nor should it be needlessly taken from its place. The cage should be provided with a drawer, which must be kept clean and strewn daily with fine gravel or sand. Cleanliness is very necessary.

Fresh water, both for drinking and bathing, should be supplied daily. Plenty of nutritious food should be given. Feed the egg and cracker mixture daily in addition to the seed and some green food, such as lettuce or dandelion or bits of sweet apple.

As soon as the male is observed to be feeding his mate, nesting material or a ready-made nest should be given to them. In about eight days after mating the female will begin to lay, and will deposit one egg daily until the whole number are laid, seldom less than four, occasionally six or even seven. The period of incubation is fourteen days.

The male will assist in feeding the young; plenty of soft food should be supplied them. The egg and cracker mixture should be customary diet, on this they will thrive. They should also have plenty of succulent green food, such as lettuce, chickweed, etc. In about three weeks the young birds will be able to leave the nest. They will soon learn to feed themselves if plenty of soft food is kept before them, and will soon eat the birdseed. When some four weeks old the males will be noticed swelling their throats, as if attempting to warble. The birds will be in full feather when six weeks old, but soon thereafter begin to cast their body feathers, and two months may elapse before they are in perfect plumage again. During this period they should be carefully preserved from draughts, and fed the egg mixture daily, together with rape seed which has been softened in water, and a little crushed hemp seed, not forgetting greed food.

A young male's capacity to sing depends upon good breeding. He inherits this, but if he sings well it is by imitation. If you expect your young birds to become good singers you must place near them as good as songster as you can buy or borrow. A little money spent for a fine singer that can act as instructor to the young will be well repaid.

Bird fanciers in Germany put their canaries to school immediately after the moulting season is past, the birds being then about three months old. A large number of young males are placed in a half-lighted room, connected with an apartment above by an opening in the ceiling. In the upper room are placed the choicest singers that can be commanded—nightingales, larks, etc.—which act as instructors to the young birds. The young canaries soon learn the lessons so carefully set for them and in a few months become expert musicians.

These two articles on both English and German canaries should be helpful to canary breeders and either of the writers will be glad to answer questions or supply stock to enquirers.

BASLEY FORMULAS (Tested)

Basley Chick Feed

Cracked Wheat	30 lbs.
Steel Cut Oats	30 lbs.
Finely Cracked Corn	15 lbs.
Millet	10 lbs.
Rice	10 lbs.
Pearl Barley	10 lbs.
Rape Seed	10 lbs.
Granulated Milk	10 lbs.
Granulated Dried Bone	10 lbs.
Chick Grit	10 lbs.
Granulated Charcoal	5 lbs.
Total	**150 lbs.**

Basley Dry Food for Laying Hens

By measure:

Bran	2 parts
Alfalfa Meal	1 part
Corn meal	1 part
Rolled Oats or Oatmeal	1 part
Beef Scrap	1 part

A little pepper and salt.

Basley "Egg Coaxer"

Dose half a pint once a day for twenty hens when they are moulting or to encourage egg laying. This is an infallible egg producer. To be given in the mash either dry or wet.

Dried Blood	10 lbs.
Beef Meal	10 lbs.
Bone Meal	10 lbs.
Linseed Meal	5 lbs.
Sulphur	2 lbs.
Powdered Charcoal	2 lbs.
Cayenne Pepper	½ lbs.
Salt	½ lb.

Douglas Mixture

Tonic and disinfectant: Sulphate of iron (common copperas), eight ounces; sulphuric acid, one-half ounce. Put into a bottle or jug one gallon of water; into this put the sulphate of iron. As soon as the iron is dissolved, add the acid. When the mixture is clear, it is ready for use. Dose: one teaspoonful in one pint of drinking water. This is one of the best tonics for poultry known. It is an antiseptic as well as a tonic, and is a good remedy for many diseases.

Basley Liniment for Rheumatism

One cup of vinegar; one cup of turpentine; as much saltpetre as it will take up, about a heaping tablespoonful. Keep in a bottle, shake before using. Bathe the affected part twice a day. Excellent for bruises, sprains, etc.; also in the human family or animals of any kind.

Epsom Salts, Purgative Dose

Epsom salts is one of the most useful drugs we have in combating internal diseases in poultry. An ordinary dose is 20 to 30 grains, administered in water. The dose for different ages, where quick purgative effect is desired, follows:

Age of Bird.	Amt. per Bird.	How administered.
1 to 6 weeks	10 grains	In feed
5 to 10 weeks	15 grains	In feed
10 to 15 weeks	20 grains	In feed
15 to 26 weeks	30 grains	Dissolved in water
6 to 12 months	35 grains	Dissolved in water
1 year and over	40-50 grains	Dissolved in water

One ounce apothecary weight is 480 grains. One ounce is a quick purgative dose for 12 mature fowls. An ordinary dose is half this quantity.

BROODERS

Will you tell a beginner what kind of brooder you recommend?—Mrs. J. F. Y.

Answer.—There are a number of good brooders on the market. For a beginner I usually advise the kind that bring in fresh warm air. Or else a small house which can be used as a coop for the chickens when they are half grown. There is a very good brooder made here which has a coal oil heater at the back that warms a small hover inside and does well in the coldest weather for about a hundred chickens and the same house or coop can be used by putting in perches when the chicks are old enough to be weaned from the hover.

At Petaluma many large breeders are taking out all the different pipes for heating and substituting a small stove heated by distillate. This stove stands in the middle or the house, which is fourteen or even twenty feet square, or about that. Over the stove is a deflector, shaped like a Chinese umbrella, which deflects the heat down upon the chicks which spread around on the floor. This brooder house and stove is intended to hold from a thousand to fifteen hundred chicks and for those intending to raise large numbers this seems to be the best and the newest way. The stove, with automatic regulator to control the heat by shutting off part of the distillate and making a smaller flame, can be bought, ready to put up, with tank, pipes, etc., at Petaluma for about $18. I have seen it working in several large ranches. (See illustrations page 75 of Mr. Davison's brooder houses.) I can recommend these. Also the fireless brooders which can be used in small, low colony houses, or even in piano box coops very advantageously for a few chickens, twenty-five to a hundred.

PART II.

Questions

and

Answers

CAUSE AND CURE OF SICKNESS

Apoplexy—What is the trouble with my hens? They seem healthy and all at once they begin to gasp and fall over dead. I cut one open and it was in fine condition, fat and nice. I cannot make out what it is.—Mrs. C. S.

Answer — Your hen had apoplexy from being overfat. The overfat condition weakens the muscles, and the heart and brain give way. Give the whole flock a little Epsom salts in the water for a week. cut down the amount of grain, especially any corn or corn meal in their feed, and feed more green food and more animal food with, of course, charcoal and grit.

Air Puff—Barred Rock about 6 or 7 weeks old. A few days ago it went to limping and I supposed it was some of the others crowding, but I have since noticed its whole right side was puffed away out, just the skin, and I took a needle and made a small opening and there was nothing but wind in it. I repeated the same operation next day. It eats and drinks and aside from the limping, seems to feel all right.—Mrs. J. N. H.

Answer — Your chick had what is called "Air Puff," and you did just right in puncturing the skin; you saved its life by it. The trouble comes from a wound or abrasion of the lung tissue resulting from violence of some kind. After caponizing a chick this trouble often develops. I have seen the poor little things almost as round as a ball and so light from the air under the skin that the slightest breeze rolled them along. Chicks that get trampled on by their mothers, or cockerels that fight, are liable to suffer from injuries that result in "air puff." They become inflated with air. The treatment is a good nourishing diet. I resort to bread and milk in such cases. It is easily digested, and, puncture the skin to let the air out. In slight cases where there is only a little air under the skin it will disappear gradually without treatment, but if there is a considerable amount of air it is necessary to prick the skin and let it out.

Bumble-foot—I have a lame hen; she limps on her left foot. She eats as well as my other hens, her comb is red and looks healthy as the others.—Mrs. M. M. C.

Answer—Your hen has probably what is called "bumble-foot." It is something like a stone bruise or a corn in human beings. It usually comes from a corn or bruises of the feet, wounds with thorns, broken glass, hard stones or other sharp substances. The ball of the foot becomes swollen, inflamed, hot and painful. The fowl appears in pain. Corns are often caused by too small or narrow perches, which compel the fowl to grasp them tightly in order to maintain their position. This firm grasp continued night after night, affects the circulation of the part of the foot that comes in closest contact with the perch. A similar condition may be caused by heavy birds flying from their perches and lighting upon a stony surface or hard floor.

If it has not yet become an abscess, simply cut off the thickened skin or corn without causing bleeding and paint the corn with tincture of iodine. If pus has developed, soak the foot in warm water twice a day and poultice until the inflammation is reduced. After thoroughly cleaning the foot, if pus has developed, open the abscess freely with a sharp knife and scrape out the diseased matter. Wash out the wound carefully with peroxide of hydrogen or carbolized water. Stuff the wound full of iodine gauze and bandage it. Continue this treatment daily until the wound is almost healed, then apply a good ointment daily until it is entirely well. The bird must be kept on clean, dry straw until fully recovered.

Bronchitis—Will you kindly tell me what ails my White Leghorn hen? She sits around most of the time and squawks and slings her head and when I hold my ear to her side I can hear a continual rattling. Her comb is red and she eats well. I feed corn, wheat, Kaffir corn and table scraps. They run on plenty of green range. Her nostrils are clean. Age, 8 months.—C. C. S.

The irritation of the bronchial tubes is sometimes the remains of an attack of roup. I have found a little honey one of the best remedies. I would advise you to mix one teaspoonful of eucalyptus oil or teaspoonful of turpentine (I prefer the eucalyptus) in one cupful of strained honey; mix thoroughly and give the bird one teaspoonful night and morning. At the same time give a nourishing diet. A little red pepper and chopped onions in her food would also help the cure.

BALD HEADED—Some of my hens are becoming bald headed. The feathers for half an inch and more back of the comb disappear. The hens seem in the best of health and lay well. There are no lice or mites on the chickens, on the roosts or in the nests. If you can give me a remedy I shall consider it a great favor. —Mrs. E. E. C.

Answer—This is not at all an uncommon occurrence just before the moult. Those feathers have merely ripened a little earlier than the others, and, strange to say, it is usually the best layers that are so affected. You can grease the bald spot with a little vaseline. This will hasten the growth of the new feathers.

BLIND CHICKS—What is the matter with my little chickens? They are about two months old. I find them with one eye shut and sometimes both, and when I open it a watery substance comes from them. When only one eye is affected, they are perfectly blind in it, but can see all right out of the other, and when both eyes are affected, they are blind in both.

Their mouths are perfectly clear and they have a rattle in their throat. They have been affected now for about two weeks and several have died. It seems very contagious.—Mrs. A. L. S.

Answer—The starting point of nearly all cases of blindness in chicks is in roupy breeding stock. A slight chill or cold is sufficient to start an epidemic of this blindness in a flock of chicks, if they already possess the inherited tendency to weakness of these parts from parents that were not in fit breeding condition. This blindness is a result of an inflammation of the mucous membrane of the eye and lids, which produces a sticky exudate, which gums the eyelids together.

Sometimes the inflammation of the lids is excited by irritating substances like lime or sharp, dusty sand, insect powders or kerosene getting into the eyes. These causes may produce blindness in chicks that do not have roupy ancestors. That form of inflammation of the lids accompanied by hardening of the lids is not uncommonly caused by irritants, kerosene particularly.

Uncleanliness is another cause of blindness of this sort, and too many who attempt to raise chicks are careless in this respect. Lice and mites also do their share to cause the trouble.

The best way to remedy such cases is to prevent them or remove the cause if possible. In cases where there is an amount of exudate it will be well to bathe the eyes with a solution of boracic acid, fifteen grains to a half cup of water, and then dry with a soft cloth and apply a little carbolic salve. It is difficult to get satisfactory results dosing young chickens with medicine, but you might give them either a little bread and milk with a sprinkling of red pepper and sulphur on it, or rice boiled in milk with a tablespoonful of ground cinnamon for each pint of milk.

CANCER—The writer wishes to know if poultry are subject to cancer.—J. H.

Answer—Poultry are not subject to cancer, but they are to tuberculosis, which may be taken for the same. There is no cure for this but the hatchet. A thorough disinfecting of the premises must be made. The bodies of any fowl dying from this disease should be burned, or buried very deeply, as it is an infectious disease.

CANKER—I am anxious to know if the heavy Black Orpingtons are hardy. I have just bought a fine cockerel and four hens; one of them has just got canker. What is the cause and remedy? —Mrs. M. N.

Answer—The Black Orpingtons are very hardy. Am sorry your pen has canker. The cure for that is to paint the spots with sulpho-carbolate of zinc (four grains in an ounce of distilled water) night and morning. This will kill the germ, but in case it is diphtheritic roup, would advise you to paint it one day with the sulpho-carbolate of zinc and the next day with peroxide of hydrogen, as the latter kills the diphtheritic germ. The open front houses are the best for every kind of fowl in this climate. A change of diet will often affect the droppings of the fowls, when they are normal. You had better slightly change the foods, or if you feed them charcoal, it will materially assist the digestion, and you need fear no trouble. A little Epsom salts in the water, if the fowls are very fat and heavy, is also an assistant, but by giving them plenty of green food, you will have no trouble.

CANNIBALISM—I had a hatching of Black Minorcas three weeks ago of 115 chicks; today I have about 80. In the first place, the chicks are hearty and well, but will bite the rectum of the other chicks and in two or three minutes will just tear the bowels out and kill the little chicks. Every one will give it a nip, and if we are not constantly on the alert all would be dead. No one of whom I have inquired has ever heard of such a thing. I have raised these just as I raise my White Leghorns. I hatched 160 seven weeks ago, and today have 158 fine chicks. You would oblige me very much with a remedy.—W. P. H.

Answer—The remedy for "cannibalism" is first, to keep all the chicks busy with exercising; in order to do this, keep the floor of the brooder covered with chaff or finely cut alfalfa hay at least an inch deep and feed the chicks small grain (chick feed) in this; the hay or chaff keeps the toes and feet covered, conceals them, and the busy little things are so occupied scratching that they do not get into mischief. Secondly, give them a little more animal food or milk. The cannibals have a craving for animal food, and sometimes a bit of fat salt pork, whether fed to them or nailed up where they can peck at it, satisfies this craving. Thirdly, find the first leader of this mischief, and either kill him or isolate him and give him to a hen to bring up. This bad habit is usually started by one chick, and all the others follow suit, and soon the whole brooder will acquire the habit, and it is almost impossible to stop it if it has got a good start.

WARTS ON COMBS AND EYES—I am in trouble and I know you can advise me. September 24th I hatched some Blue Andalusians. They have grown very fast, seemed extra healthy and vigorous until a few days ago, when warts began to appear on their combs and eyes. In one night they grew twice in size. I have nine, and they are all becoming affected. What in the world is it, and is it catching? They have run at large entirely and their feed in grain is mostly kaffir corn. They were such fine chicks, and I was raising them for breeders, but now feel discouraged. I have a younger litter, four weeks old, but they are all right so far. My old birds are fine stock and very healthy. These warts did not make their appearance until the chicks were eight weeks old.—Mrs. H. E. S.

Answer—Your chickens have chicken pox in a very virulent form. Chicken pox is from a germ and it is very infectious. It is fatal to young chicks. In severe cases it goes into the throat and mouth, as you describe. The best home remedies that I know are first to grease the "warts" that are on the outside of the mouth or under the wings with a little carbolic salve. Then wash the mouth and throat with vinegar and salt (a level teaspoonful in a cup of vinegar), following this the next day with swabbing with peroxide of hydrogen. Give germazone in the drinking water. Feed nourishing and easily digestible food, such as bread and milk.

Chicken pox or sore head affects ordinary fowls, and more rarely geese. Young chickens are more susceptible than the older fowls. It is caused by a parasitic fungus. The bite of insect abrasions of the comb, such as scratches from fighting in cockerels or turkeys, make conditions favorable for the parasite to get into the skin of the fowl. The bite of an insect, such as the flea or mite, will carry or give the disease. It is contagious. At first it has some appearance of warts, these reach their full development in from five to ten days. The largest are found about the beak, nostrils or eyelids. These warts seem to run together and form yellow masses upon the comb and wattels. Chicken pox is more prevalent in damp weather than in dry.

The cure is, wash the warts in warm soap suds, dry and apply carbolated vaseline, or kileroup, and feed a light, nourishing diet, one-third being cut green alfalfa and give bread and milk to which add half a teaspoonful of powdered sulphur.

Disinfect the premises thoroughly.

COLD IN THE HEAD—Can you tell me what is the matter with my chickens? They eat, seem to feel good, sing and play and are laying good, but they seem to have a cold or something. They try to blow their noses and bubbles come out. Have been that way for about six weeks; they have a good coop with no air holes; six eight; one end open; only twenty-five to roost in it. They have had bluestone in their drinking water every day for a month; they do not get any worse or seem to be any better; they have warm mash for morning feed and wheat noon and night.—F. C. H.

Answer—I am afraid that your chickens are too crowded in their roosting quarters and that they get too warm at night and come out into the cool morning air and in this way take cold. Or the open end may be towards the night breeze. They evidently have, for some cause, slight colds. Bluestone, or germazone in the water is an excellent cure and by adding chopped onions and a little red pepper to the mash, should cure them. One teaspoonful of red pepper for every twelve hens is the dose. Rub the head well with kileroup. See roup cures in this book.

COUGH AND SNEEZE—Will you please tell me what is the matter with my birds? I have several that cough or sneeze, I do not know which. They will shake their heads and "holler." One can hear them quite a distance. Will you please tell me the disease and remedy?—B. J., Tucson, Ariz.

Answer—Your fowls have bronchitis and perhaps some influenza. Give them bread and milk for supper, and a quinine pill and half a teaspoonful of red pepper mixed with butter. And see that they do not sleep in a draught or in a house where the rain comes in on them. Also give them each five drops of eucalyptus oil on a bit of bread or in half a teaspoonful of honey.

COMB DISCOLORED — I have a White Leghorn cock two years old; he has always been healthy, but for the last two months I notice that his comb and wattles turned a deep purple and would remain so for days, then they would change to a natural color again, but only for a day or so, and then turn purple again. He seems to be healthy and vigorous in every way. Now, can you tell me what can be the matter with him and what I can do for him, or if it would be wise to use him any further for breeding purposes?—Mrs. L. S.

Answer—The comb tells quite a little story of what is going on in the organs of the whole body. Any change in the appearance of the comb is indicative of a disturbance in some other part of the bird.

The dark colored comb is an indication of a disordered liver and indiges-tion. The dark comb is one of the first symptoms noticed in congestion of the liver and most cases of this come from an overfeeding of a ration too rich in starch elements, such as too much potatoes or bread in the table scraps, and insufficient exercise. I do not know how you are feeding your fowls, but I would recommend you to put a little Epsom salts into the drinking water, or you can give him alone a small half teaspoonful in a tablespoonful of water, and put in the drinking water of the whole flock ten drops of tincture of nux vomica to a pint of water. Feed plenty of green food and more meat than you are now giving; keep this up for a week and then turn the birds out on a grass range if possible, otherwise give to the birds as scratching material the waste from an alfalfa hay mow and allow them only a little grain, wheat, and make them scratch hard for that. It would not be advisable to use the male bird for breeding. Breed only from the most vigorous stock you have.

WHY COMBS ARE WHITE—We have two Buff Orpington hens that are sick. They mope around and do not eat. Their heads and gills are almost white, and sometimes one is almost blue. They look as though they have lice, but they have not. Can you give me some advice as to how to treat them? Thanking you in advance, I am, respectfully, A. G. O.

Answer—The comb tells quite a little story as to what is going on in the organs of the whole body. The normal condition of the comb presents a healthy look that the poultrymen call the "standard red." Any deviation from this red is an indication of changed action in the workings of the organ, or to a change in the vitality of the whole bird. The light colored comb shows an anemic state of the bird and is a sign of underfeeding, lice, poor ventilation, and absence of green vegetable food, impure water and uncleanly surroundings.

As you say nothing of the feeding and treatment of the birds, I am unable to say which of these conditions fits your case. I think probably they are infested with lice or their houses with mites, and the only remedy is the extermination of these.

CATARRH—Can you please tell me what the trouble is when chickens cough and their nose runs, also state the best way to rid them of this plague?—Mrs. S. A. B.

Answer—Your chickens have taken cold and may have lice. Try to discover what is giving them their severe colds. It is probably some draught. Put a piece of bluestone in their drinking water (the size of a bean in a quart of water) and give them a pill of the following: Mix two tablespoons of lard, one each of mustard, red pepper, vinegar; mix thoroughly, add sufficient flour to make a stiff dough. Give a bolus of this as big as the first joint of your little finger every night. One or two doses usually cures.

CROP BOUND—I have about 100 Leghorns; been very healthy all winter; laying good. Now about six weeks ago I lost eleven of the heaviest ones in six days. They had yellow droppings; lived only two days and died. Four others died after having a heavy crop hanging down; they were apparently healthy and laying eggs regularly; I cut the crops off three of them and found nothing but long strings of hay. Please oblige me by telling me the cause and what remedies.—A. F. H.

Answer—Your hens are suffering from what is called crop-bound. They eat long pieces of hay, which form into a ball in the crop and cannot pass through them. After a time this ferments and decays and poisons the chickens or brings on inflammation of the crop.

Cases of impaction of the crop caused by cracked corn are not so common, and occasionally there is a case from some foreign substance filling the outlet of the crop.

Treatment—Make the bird swallow, by the aid of a funnel, some water in which half a teaspoonful of baking soda has been dissolved, then work the crop with the fingers until soft. Turn the bird upside down, and by working the crop, if it is impacted grain, the mass will be vomited out. After treating the bird, give it at night a dose of castor oil and feed sparingly for a few days on soft food.

When long pieces of grass or hay cause this trouble, as in your case, almost the only remedy is to cut open the crop of the bird and wash it out. Have someone hold the bird so you can have both hands free to work. Pluck enough feathers from the breast to give bare skin half an inch wide by two inches long. Then with a sharp knife cut through the skin, lengthwise of the bird, an opening one inch long over the place of the swollen crop. Cut only the skin, leaving the crop untouched until the blood of the first incision has ceased to flow. Then cut through the crop a little over a half inch long. Half an inch may seem short, but you will be surprised to see how large the opening is after you have worked through it for awhile. In removing substances from the crop, be careful to let as little as possible slip between the skin and crop; with a button-hook or anything else handy, remove the contents. If filled with grass or hay, it is sometimes necessary to cut the mass with scissors before any start can be made. When the crop is apparently empty, push your little finger into it, feeling to know whether there is any obstruction at the outlet. If you find the opening clear, the last thing is to sew up the cut. With needle and white silk thread. take two single stitches in the cut in the crop, then in the same way take three stitches in the skin, tying off the silk at each stitch. Be careful not to include the crop in the knot tied. After the operation feed soft food, omitting grain for a week.

SICK CHICKS—I want your advice. My little chicks seem to be pert and healthy when they are first hatched and all right until they are two weeks old, and then they get all pasted up in the back; don't eat, just drink and are sleepy looking, droopy and die. I have lost over a dozen that way and have a lot more now that are in the same condition. They have no lice or mites, for I have examined them, and I don't see how they take cold. I have barrels for them to roost in, with a screen in front to protect them from cats or rats, so there is no draught through the barrel and I don't feed them anything but chick feed. I put copperas in their water this morning to see if that would check it. I am sorry to lose all my chicks after I have taken such good care of them. Please let me know as soon as possible what I can do for them and oblige. Yours truly.—Mrs. C. C. B.

Answer—Your little chicks have taken cold, probably from sleeping in a barrel. When little chicks have bowel trouble, it is almost always from taking cold. In mature hens a cold affects the head, throat, bronchial tubes or lungs, whilst with little chicks it affects first the bowels.

A fireless brooder might have saved all your chicks. A barrel is very cold, unless it is well banked up on the outside and the nest inside very carefully made. A flat box is much better. Copperas will not help them; the best thing for them is rice, boiled in milk, with a tablespoonful of ground cinnamon to each pint of the milk added after cooking. Cinnamon is a good disinfectant and healing and warming to the bowels. Copperas is cold and chilling and is apt to give indigestion to small chicks.

PULLETS DYING—We have a flock of incubator chicks that are not doing very well. The little pullets started to die when but seven weeks old and we lose one or two every day. They have the whole farm to run on. At first they hang their wings and act sleepy, then their heads turn blue and they die. We cannot find lice nor fleas on them. They are fed wheat, oatmeal, and some onions and malk. Have plenty of water, grit and charcoal.—Mrs.T. L.

Answer—I think your chickens have worms; the wings dropping and their acting sleepy are two of the most prominent symptoms with worms. Cut open the next one that dies and examine it. The best cure that I have found for worms is ten drops of turpentine in a teaspoonful of castor oil. This is for the common round worms. For tape worms, which are not so common, the dose is ten drops of tincture of male fern on a piece of bread or a lump of sugar in the morning, fasting followed by a dose of castor oil in an hour. Be careful to clean up and destroy the droppings or the other chickens will eat them and the trouble will increase.

DIPHTHERITIC ROUP—Having derived many useful ideas from your writings, I take the liberty to ask your advice regarding a disease which has come upon my chickens. The first symptoms seem to be a sneezing or squawking sound, as if the chicken had a beard in its throat; then a white membrane forms over the windpipe and the eyes close up and lumps break out around the comb. The lumps finally break and the eyes and nose run. Both Barred Rocks and White Leghorns are afflicted. The Barred seem to suffer the most.—Mrs. R. F.

Answer—I am sorry to say your fowls have diphtheritic roup. It is a very infectious disease and if you have children you had better keep them away from the fowls. Spray the mouth, throat, nostrils and cleft in the mouth twice a day with peroxide of hydrogen. Give the fowls a quinine pill, four nights in succession, and once a day a bolus of the following mixture: Two spoons of lard, one each of mustard, cayenne pepper and vinegar; mix thoroughly, add flour enough to make stiff dough; give a bolus as large as the first joint of your little finger once every twenty-four hours.

FATTY DEGENERATION OF LIVER—I have noticed a hen moping and eating but little for two or three weeks, but as I had broken some up from sitting, thought it the result from broodiness. However, as she got no better I separated her from the others, but yesterday she died. This morning I did as you advised, and duly performed the autopsy. I saw at once on making an incision what was the matter. Her liver was so enlarged that it occupied almost the whole cavity. I never saw one such a size. It was covered in blotches of pink spots, small as a pin point. There was fat around the heart and the intestines; perhaps a fifth of an inch thick. There was plenty of grit in the gizzard but no food. The heart seemed in good condition, the body a good color, and flesh firm. In the cavities of the back is a substance, of which I do not know the name, that seems to be enlarging and hardened. There were many eggs, but very small and undeveloped. Is this the kind of liver which is used as a delicacy and produced by overfeeding? My fowls were fed corn all winter and were much too fat this spring. In March they had layers of fat an inch in thickness. I did not suppose that a laying hen ought to have any fat inside of her. How should that be?—G. S. H.

Answer—Your hens certainly had fatty degeneration of the liver, or the disease which the overfat geese have when their liver is considered a delicacy. She simply had been fed an unbalanced ration containing too much of the fat element, and being a Plymouth Rock, had become overfat. The substance in the cavities of the back is the kidneys. There are three lobes of these on each side. Your fattening ration had also affected them. So much fat will also affect the egg laying, will make small eggs and chickens will be weakly, as there will be preponderance of fat in the eggs from which they are hatched. A laying hen should not be anything like as fat as those you describe.

FEATHER PULLING—Will you kindly tell me the cause of chickens pulling feathers from each other and eating them? We feed them wheat, cracked corn, etc., also ground bone.—G. H. T.

Answer—Various causes have been assigned for this habit, the most probable being improper rations and idleness. In some instances it is caused by mites or lice. As in some cases, the habit is due to insufficient animal matter in the rations, or to feeding too long on a single kind of grain, particularly corn, one of the first measures adopted should be a well-balanced ration, containing skim milk, meat, bone, vegetables or green feed and frequently varied. The Geneva, New York, experiment station applied to the feathers lard or vaseline in which powdered aloes had been mixed. After continuing this treatment for some time the habit disappeared, due to the disagreeable taste of the aloes. The skin and feathers should be carefully examined for lice and mites and if these are found the remedies recommended for such parasite should be applied.

HEART TROUBLE—I have a very fine rooster, two years old. For the past two months he has been troubled by some difficulty in breathing. At times his comb and wattles become purple for two or three minutes, then the color gets red again. I have looked for canker but cannot find anything that seems wrong. Have used vaseline but it has not done any good. It seems to me more like asthma or bronchitis. Wish I could cure him for he is a valuable bird.—Mrs. I. G.

Answer—I am sorry to say that your bird has heart trouble. This has been brought on by some great excitement, such as fighting, fright or being chased. It may possibly be fat on the heart, which weakens that useful organ. You might try giving him in the drinking water nux vomica and sulphur comp. 2x twelve tablets to each pint of drinking water. Be careful to give him plenty of green food and grit, besides his ordinary food. Cases of this kind are al-most incurable, but the treatment I have indicated may help him and prolong his life.

HEMORRHAGE OF OVIDUCT—I wish a little information in regard to a Leghorn hen that died yesterday. She apparently choked to death; made a queer noise. We opened her and found at the bottom of her egg bag a large clot of black blood. Can you tell me what it was and if there is any cure for it?

Answer—Your White Leghorn hen had a hemorrhage of the oviduct; this is excited by any of the causes which lead to congestion and inflammation and may be counteracted by green feed and the suppression of egg foods, stimulants, red pepper, etc. It sometimes occurs from trying to pass too large an egg. There is no cure that I know of, as death occurs before one finds out what is the matter.

INDIGESTION AND LIVER COMPLAINT—My hens are on a strike, and their faces and combs are becoming pale or yellow. What is it?—I. S. B.

Answer—You have been overfeeding, and now your fowls have indigestion. Indigestion in fowls is the cause of many ailments. With your birds it has been brought on by lack of grit, with not sufficient roughness (or filling) and too little exercise. How can indigestion be prevented? By dieting. Feed more bulky foods, such as alfalfa, and less solids. A continued grain diet of wheat, corn, barley, if few in quantities and not varied by bulky foods, vegetables, etc., will bring on indigestion, especially when but little exercise is taken. An insufficiency of clean water is also conducive to this trouble. Clover, alfalfa, any of the green stuffs or vegetables, usually fed to fowls, are absolutely necessary preservatives of health. Now, as to a remedy: Your fowls' indigestion has taken the phase of biliousness. Give each affected hen one of Carter's Little Liver Pills, and give the whole flock a teaspoonful of baking soda in a quart of water every day for a week. Give no other water. Why do I recommend soda? Because it helps to emulsify the too much fat in the bowels. You might give a teaspoonful of Epsom salts in the water for a week, to carry off the bile which is overflowing into the intestines and being taken into the system. It is not kindness to feed your fowls every time they come near you. It is far kinder to keep them working for it and so keep them healthy.

INFLAMMATION OF THE CROP—I have a Buff Orpington hen that has a disease I have never seen before. Her craw is swollen to several times its normal size and is filled with wind or gas. She eats but not as much as she should, and is getting thinner all the time.—H. Y.

Answer—Your hen is suffering from inflammation of the crop. This is like a very severe attack of indigestion. The causes of this are irregular feeding or too much food being taken at one time. Partially decomposed meat, or putrid food of any kind will also cause congestion and fermentation of the contents of the crop. The same disease occurs when birds eat substances containing phosphorus or arsenic, or rat poison. The feeding of too large a quantity of pepper or stimulating "egg food" in the mash will also cause inflamed crop as well as trouble with the egg function.

Treatment—A clean, dry pen should be provided for the affected bird. Empty the crop of its irritating and decomposing contents by careful pressure and manipulation while the bird is held with its head downward. When the crop is freed of its contents, give two grains of subnitrate of bismuth and one-half grain of bicarbonate of soda in a teaspoon of water. The bird should then be kept without feed for eighteen hours and then fed sparingly upon easily digested food, such as bread and milk. Half a grain of quinine morning and night for two or three days will complete the cure.

INFLUENZA—I am in trouble with my chickens. Five of them have died since Monday. They open their mouths and gasp for breath and sneeze and their eyes are very watery. I feed wheat, cracked corn, plenty of green stuff and table scraps, and they have a good run. I always wash out their drinking pans and rake out under their roosts at least every other morning.—Mrs. J. F. S.

Answer—Your chickens have influenza. They are taking cold in some way. Either there is a draught in their house or the rain comes in on them; a few have had the cold and they are giving it to the rest. Keep bluestone in their water, and give each of them a bolus of the following, night and morning: Mix two tablespoons of lard, one tablespoon each of cayenne pepper, mustard, vinegar; mix thoroughly, add enough flour to make stiff dough; roll out; give a bolus as large as the end of your little finger. Put carbolated vaseline or Kileroup up their nostrils and in the cleft of the mouth, and give them chopped onions in their food.

LEG WEAKNESS—I am in trouble over my White Rock chickens. I only have a few, so would like to save them. When they are about three weeks old they get weak in the legs, and after a week or so they begin to tremble like a person that is nervous. They eat well until the last. I feed boiled egg and bread crumbs. They have green barley to run on. I feed kaffir corn at night. During the day I feed onions and table scraps. If you could tell me what to do I would be a thousand times obliged.—Mrs. W. K.

Answer—Your chickens are suffering from what is called "leg weakness." Leg weakness comes chiefly from wrong feeding, also from overcrowding at night and overheating.

Young chickens should either be allowed free range with a hen or be encouraged to work and scratch for their food. This strengthens their legs. The green food should form at least one-third of their diet and for such young chickens it would have to be chopped up finely. They cannot peck off sufficient green barley. It soon becomes too tough for them. The cure for leg weakness is a little tonic (a few drops of iron in their drinking water) and plenty of green food and cracked wheat instead of kaffir corn. If it comes from overcrowding or overheating, either under a hen or in a brooder, you must rectify this. See that they have "chick grit and charcoal."

ACUTE INDIGESTION—I am in trouble with some incubator chicks and I write to ask you to be kind enough to diagnose it.

The chicks are Black Minorcas and are fourteen days old. They seemed to be doing well till yesterday. One or two all at once got so they could not stand up or walk, but looked bright. This morning there are half a dozen affected the same way. I feed them a chick feed I have used for several years, curd, charcoal, and plenty of grit and always give the fresh water three or four times a day. For the last three days they have run in a lettuce patch part of the day. I have a hot air brooder, plenty of fresh air at night. No sign of lice and I use a powder in the brooder once a week. I have raised chickens for several years but have never had any trouble like this and I would be greatly obliged if you can diagnose the case and give a remedy. —Mrs. P. V. M., Sacramento.

Answer—The symptoms you describe are those of poisoning or sudden and acute digestion. I can only suggest that it may be that the chick feed has mouldy grain in it or there may be ptomaine poison in the beef scrap. I would suggest that you put a little bicarbonate of soda in the drinking water. Give all the succulent green food that you can persuade them to eat and to each affected chick administer without delay ten drops of castor oil. Try to find out where the poison comes from, change all the bedding in the brooder and brooder house and scald the brooder thoroughly with hot soap suds. When any sudden trouble like this comes, try to find the cause of it and remove it. I feel sure it is poison of some kind, either ptomaine or fungoid, such as mouldy bread or mildewed grain.

LIMBER NECK—We have between 200 and 300 chicks two months old that are badly afflicted with limber neck, and we cannot find out the cause. The first two or three weeks we fed them millet and Johnny cake, made stiff and dry, of coarse corn meal, but they began to get sick, so changed to dry food, consisting of cracked wheat, millet, beef scraps and grit, but the chicks got no better, so now we are using just wheat and grit. They have lettuce every day and often young vegetables—tops and all. Until about a week ago they were kept by themselves in wire pens, but as an experiment my husband let them out to run, and still they get sick. They do not all die, as I bring them to the house as soon as we find the sick ones, but from one to seven die nearly every day. They have fresh water every morning. I do not try to doctor them, but just keep them warm. I have saved some pretty sick ones in that way. They are such a bother, and we have lost so many in that way. The flock which is the most affected had a habit of huddling when they were small, until they would sweat and sometimes die. Do you suppose that could have anything to do with the present troubles?—Mrs. F. L.

Answer—Limber neck is due to a disorder of the nervous system and is usually the result of disturbances of the digestive organs from severe attacks of indigestion or from infestation with worm parasites. Chicks are sometimes affected in this manner by unusually hot days and nights. I think very probably their digestive organs were weakened by being overheated when they huddled, and I would give the whole flock plenty of charcoal to eat, with plenty of green food and animal food, and no millet, as millet is very hard to digest. Give the sick birds a small piece of gum asafoetida, about the size of a green pea. Repeat the dose the second day. This will usually cure. Feed them with bruised garlic or with chopped up onions. Give them grit or very coarse sand in boxes to assist in the digestion, and I think you will have no further trouble.

It is possible that your chickens have worms. You had better open the next one that dies and examine it, and if you find it infected, give the others turpentine in the drinking water, half a teaspoonful to a pint of water (giving no other drinking water) or if you prefer it, give a teaspoonful of castor oil with ten drops of turpentine in it to each sick chick. The chickens dislike the turpentine in the water, but it will kill the common round worms if continued for a week.

LIVER TROUBLES OR POISON—I want your advice and a remedy for my sick fowls. The symptoms are briefly stated: Grown chickens affected droop for two days, comb turns black and they die. Have lost nine in two days.

My chickens have free range, fresh water and plenty of barnyard scratching with Egyptian corn every night.—C. V. N.

Answer—The symptoms you describe denote either liver trouble or poison. In your case I think perhaps it is poison, either from rat poison, gopher or some poisonous weed. You had better hold a post mortem examination on the next one that dies and then you will be able to tell just what the trouble is.

LIVER DISEASE—The liver is the largest and most important organ in the fowl's body. It not only prepares the bile which is poured into the intestines to assist digestion, but it acts also as a germ destroyer, and assists in some of the necessary chemical changes which take place in the blood. This organ contains numerous blood vessels and through it passes a large quantity of blood. It is particularly subject to the attacks of various kinds of parasites. Most of these parasites probably find their way to the liver through the blood channels, lodge in the minute blood vessels and multiply there. It is especially liable to congestion, which frequently occurs from errors in feeding, or other causes of intestinal irritation.

CONGESTION OF THE LIVER—A lack of exercise, combined with overfeeding, is the most frequent cause of congestion of the liver. It also results from the overuse of stimulating condiments and the persistent feeding of many of the so-called "egg foods" to birds closely housed and yarded. Most of the cases of liver trouble are, however, due to the overfeeding of a ration too rich in starch elements, such as too large an amount of potatoes or bread.

The early symptoms of congested liver are not always recognized, as the bird's condition may not be suspected. There is at first a lack of color in the bird's comb and wattles, followed by a watery diarrhoea, dark at first, but changing to yellow. The plumage is rough and dull. Then the color of comb and wattles begins to change to a dark red or purple, often becoming nearly black. The fowl is usually fat at this stage.

Treatment—If the early symptoms are noted and properly treated, most cases will recover. As the cause is largely one of ill-balanced rations and feeding, with insufficient exercise, a change must be made in this. Give twelve tablets of nux vomica and sulphur comp. 2x in each pint of drinking water. Feed plenty of fresh green stuff and some cooked meat. Keep up this treatment for a week, then turn the bird out in a grass range if possible; otherwise give the birds as scratching material the waste from an alfalfa haymow and allow them only a little grain (wheat) and make them scratch hard for that.

INFLAMMATION OF THE LIVER—Inflammation of the liver is really the stage following congestion. The causes are the same, and the symptoms also, only increased in every way. There is little satisfaction in treating a case that has drifted into inflammation. The liver tissues are permanently injured and that organ is unfit to perform its duties.

Treatment—Treat these cases, if at all, by clearing out the bowels with a dose of castor oil or sulphate of magnesia, following this by nux vomica, as in the congestion of the liver. Feed lightly, depending upon bran and clover with a little cooked meat and a free range of grass.

NAKED CHICKS — Thinking perhaps you can help us, I will ask you for a little of your time. Late in October we bought a hen caring for thirty chicks. We have fed them cracked corn, meat scraps, plenty of green stuff, charcoal and grit. They feathered out, but since many of them have become bald, and the feathers fall from their neck and they are growing thin, still their wing feathers are long, making them look very queer. They are not incubator chicks, and we have examined them closely for mites, have dusted them for lice and they are quite free from either. What do you think is the cause and what can we do for them?—H. A. S.

Answer—Your chickens are huddling at night, crowding too closely together. This makes them sweat and their feathers fall out. Put a little carbolated vaseline on their heads and cut the feathers of their wings as close as you can without making them bleed. Give them wheat and more meat in their food and try to prevent their crowding at night. It is the crowding and lack of wheat in the food, lack of protein, that prevents the feathers growing, and the sweating makes them fall out and will make the chickens thin.

OVARIAN TUMOR—I had a nice Orpington hen; she had been laying each day and appeared to be perfectly healthy; comb red, went around seeming quite well. I feed cracked corn and wheat, table scraps, and the chickens have good range and plenty of good food. About four days ago the Orpington appeared to be lame in the right leg. I caught her, examined the foot and leg, could see nothing wrong and she continued lame, and with difficulty got on the nest. To all appearances the leg was broken, as it was harder for her to walk each day. Rather than see her suffer I had her killed. I dissected her; she was very fat with an abundance of eggs, one soft shell. I found in the right side of the back a growth about the size of a pigeon egg, which appeared to be part of the egg bag. The liver and other organs appeared to be healthy. I hope that you may be able to tell me what the growth was and if there is a cure for it, in case any of the other hens have such symptoms. The hen was about two and a half years old. Would age have a tendency to hinder her?—Mrs. H. R. B.

Answer—Your hen had what is called an ovarian tumor. The trouble is very common, and yet we don't know very much about it. I am inclined to think that if investigations covering a large number of fowls kept under a variety of conditions were made, it would be found that cases of tumor like this are more abundant among fowls kept closely confined, or fed heavily for egg production, than among those kept under more natural conditions. It is quite reasonable also to suppose that the offspring of hens heavily forced for egg production would show weakness of the reproductive system, resulting in diseases of this character. It possibly also may come from an injury of some kind. Undoubtedly some strains or families are more subject to it than others. There is no cure for it and the only preventive is to keep the hens healthy and busy.

To Remove Spurs — The English method of removing spurs: to successfully remove spurs from an old bird it is necessary to have a couple of baked potatoes fresh from the oven; first take a long strip of damp rag, wind this around the bird's shank, both under and above the spur, so that the potato will not burn the bird's leg. Next take one of the hot potatoes and place it upon the spur, driving the spur well home. Allow the hot potato to remain upon the spur for five or six minutes, then remove the potato and with a sharp pen knife nick around the base of the spur, then insert the point of the knife in the point of the spur and gently pull, when the whole of the outer shell will come away. The next thing to do is to shave off the point of the remaining spur and the bird will be much better and safer for breeding purposes.

This English plan does not appeal to me, but as it is a novelty I give it.

The Eastern way of removing spurs, or young calves' horns: Take vaseline and oil around the root of the spur, take a stick of caustic, moisten and rub the points thoroughly. This has to be done when the birds are young, say six months old. The calves' horns are re-

moved in the same way by clipping the hair off and rubbing the small horns. This has to be done when the calf is only a few days old.

I have always removed the spurs by sawing them off and then filing down the rough edges.

Overfat Hens—I have about two dozen Buff Orpington hens and have had no eggs for four months. They appear as healthy as can be. For some time I fed them wheat twice a day and the table scraps. I began to think I was not feeding the proper foods; then I got bran and an egg maker and also bought cabbage for them and still no eggs. They have lots of exercise and gravel and are so fat you cannot eat them. Please tell me what to do to reduce the fat. The past two weeks I have been giving them just the scraps from the table. Tell me, is that the proper method to reduce fat?—Mrs. A. C. S.

Answer—Your hens are so fat that they cannot lay. The whole inside of them is filled full of fat so the eggs cannot pass down the egg duct. The best plan would be to kill and eat, or sell the fowls, because they will not make satisfactory layers after being so fat.

However, if you wish to keep them, your only plan will be not to give any grain, or any table scraps until they are reduced in fat; give only green alfalfa or lawn clippings, for two weeks, then commence and feed half an ounce of meat per hen per day and lawn clippings; no grain or bread, and in about a month they may begin to lay.

Pendulous Crop—I have a hen, and its crop hangs down so far that when it walks its feet are always hitting it. We cut it open once and only the corn and feed it had eaten came out of it. I have thought I would kill it, but I was afraid it might be a tumor and that the hen would not be fit to eat. She seems healthy otherwise.

Answer—Your hen has a pendulous crop. This is usually caused by overfeeding of mash at some time in her life. It sometimes can be cured by a surgical operation. I would advise you to kill and eat the hen, as in time the crop will become sore. You can easily see before you eat it if a tumor has developed, in which case bury it.

POISONING—For some time I have read your articles and know that you are different from the majority of poultry writers, in this, that you know what you are writing about. I wish to ask you to please tell me what is ailing a fine White Wyandotte cock I have. He has been ailing about two months. He was just starting in the moult when he commenced looseness of the bowels which I cured, when one evening, as I came to shut them up, I found him on the ground unable to get on the roost; when I lifted him on the roost he fell as though dizzy and tumbled over and over. Ever since that time he has been getting worse. Now, with the least excitement, he will squat on the ground and twist his head and neck entirely around, often with his bill turned straight up.

Answer—The symptoms you describe are those of ptomaine poisoning. This is caused by bad meat or bad milk or spoilt beef scraps. Also by musty or sooty grain and formaline. The treatment is: Give a pill of asafoetida about the size of a pea every night for a week; for the same length of time put bicarbonate of soda in the water, about a teaspoonful to a quart of water; give him some charcoal in the feed and avoid feeding whatever is causing the trouble.

The preservative which butchers put on the meat acts as a poison and many fine birds have been lost by this without the owners discovering the trouble. It seems to partly paralyze the bird.

———

PTOMAINE POISON—I am in great trouble and come to you for advice. My splendid White Leghorn chickens are dying like flies and I do not know the cause nor what to do for them.

Today I lost ten and I am afraid I may lose the whole lot of them. I opened several to see if I could find the cause, but they look all right. with the exception of the crop, which has nothing in it but wind or air. The chickens are seemingly all right, and suddenly they will lie down, put their heads under their bodies, and after a while they will die.

My chickens have plenty of exercise, lots of green food, grit and running water. They can run at will all over the ranch, and I feed them some every day. I am putting some pulverized asafoetida in their mash as a disinfectant. My chicken house is new and in good order. —Mrs. K. C. Polasky.

Answer—Sudden symptoms such as you describe come from poison of some kind which brings on an attack of acute indigestion. The difficulty is to decide what the poison is and where the chickens get it.

I think your chickens, being on free range, are finding and eating putrid animal food of some kind and that they are suffering from ptomaine poison.

Rotten vegetables or moulded grain or vegetables have the same effect, although that is from a fungoid poison.

The treatment in either case would be about the same. First remove the poison from the ranch, look for any dead chicken, bird, gopher, etc., and bury deeply or burn. Continue the asafoetida in the mash, but also add a teaspoonful of castor oil for each chick the first morning, and in every mash for some time to come put powdered charcoal and sulphur, a quarter of a teaspoonful to each chick.

———

POISON—I thank you very much for your kind advice. I feed now as you direct me, with fairly good results. The beef scrap, of which I send you a sample, I bought at ———, and it killed my chickens.

I fed it to different flocks at different times with the same result and I am positive it is this beef scrap and nothing else that poisoned my chickens. I wonder how many people have lost chickens through these same people who sold to me. Perhaps they sell good scrap sometimes, but this is bad and smells bad.

What is the best way to feed rabbits to hens? I cannot grind them in a bone cutter, can I?—J. H.

Answer—The beef scrap that you sent me certainly does not smell at all good. It often occurs in the summer that beef scrap that may have been good earlier in the year has become moist or heated and a poison has developed in it, so in the summer I advise poultry raisers to buy it only in small quantities and try to have it as sweet as possible.

You know I feared it was the beef scrap and so advised you to use milk and wild game and to avoid the beef scrap. You will have to skin the rabbits or squirrels, and then you can surely grind them up in your bone cutter or if you cannot, you might hack them up with a hatchet on a block of wood, or you can boil them and let the hens peck the meat off and then chop the bones up on the block. The hens will come running when they hear that hatchet chopping. I have had them running a quarter of a mile to get the bones that were flying off the hatchet. The rabbit and squirrel bones chop very easily and the hens do love them.

POISON—I want to know what is the matter with my friend's chickens. They are a mixed flock, one year old, all laying. They are fed on scraps or garbage.

The first thing she noticed they were on the roost hanging their heads down as far as they could stretch. Then they fall on the ground and run their heads out as far as they can, and die three or four days later. She has lost seventeen.—Mrs. F.

Answer—This is what is called "limber neck," and comes from poisoning by bad (putrid) meat, fish, or garbage that is mouldy. Tell your friend to put a little bicarbonate of soda in the drinking water—a small teaspoonful to a quart— and to give also ground charcoal in the food and give each hen that is so affected a dose of either Epsom salts (half a teaspoonful) dissolved in water, or a teaspoonful of castor oil.

MILDEW POISON—Will you kindly answer the following questions: My White Leghorns are dying from bowel trouble. Two were sick for two days. I have noticed this since I began feeding a dark variety of wheat or mildewed wheat. The hens have not laid well and their combs are dark. I think it is the wheat. Will you please tell me a remedy? Do you think it is the wheat?— Mrs. J. W. H.

Answer—Mildew is poisonous to fowls and the wheat you are feeding them is killing them. Stop giving them that wheat, and give them a little charcoal in their food and also a little carbonate of soda in their drinking water, about a half-teaspoonful of bicarbonate of soda to a quart of drinking water. But there will be no use of doctoring if you keep on feeding them the poisonous wheat.

PIP—I have read your remarks carefully for over a year, but do not remember anything about pip. All my flock have it, one year and three days old. How do they get it? Is it hereditary? If so, is it in the strain or the breed, White Wyandottes? Is it fatal? If so, in what time? What is your treatment?

Thanking you for your reply, I am, very respectfully.—W. H.

Answer—I have not seen a genuine case of "pip" for many a long year—in fact, never in California. The poultry medical books here assert that it is only a symptom of a disease and not a disease at all; that it is only a dryness of the tongue produced by feverishness and rapid breathing. However, I well remember the disease at my grandmother's, in Europe, and there the cure was very simple.

The pip there was a real disease. It was a small horn or scale that grew on the end of the tongue. The tip of it was quite sharp, almost like a thorn, and the edges were almost as sharp as a knife. The sharp point and edges seem to prevent the fowls from picking up and swallowing the grain and they die of starvation.

When we noticed a hen which dropped the grain we examined her and if we found a hard, sharp scale on the tip of the tongue we would remove it with the thumb nail, scaling it off, commencing under the tip of the tongue. Then we touched the spot with borax and honey and gave the hen a dose of Epsom salts, about a quarter of a teaspoonful, or a lump of very salt butter. We fed soft food for a few days. The hens recovered quickly.

POISONED—Yesterday morning I found nine big chickens in my yard dead and about twelve more are dying. What is the cause? They sit on the ground, do not eat and the head hangs loose on the ground. The comb is dark and in the throat is a sticky slime like white mucilage. No bad smell; sometimes they jump a foot and lie down again. I fear they will all die. To a few I gave a teaspoonful of olive oil, and to some others fresh milk. I cannot imagine what it is.

Other fowls in the next yard are not affected, and all had the same food.— Mrs. F. C. P.

Answer—Your chickens have limber necks from ptomaine poisoning. Give the whole flock hypo-sulphite of soda; dissolve one teaspoonful in a quart of drinking water. And to each chicken that is affected give a piece of asafoetida about the size of a green pea. Use the gum form, and repeat the dose the second day. This disease usually comes from severe attacks of indigestion caused by eating bad animal food, or the decaying carcass of a dead animal. Putrid meat or putrid milk will cause it.

RHEUMATISM—I have a White Plymouth Rock hen about eight months old, which seems to have rheumatism. She is very fat, and a few days ago she walked lame in one leg and the next morning she was lame in both legs and now she cannot stand erect, but walks and crawls on her legs, the legs being drawn up under her so that in moving around she does not seem to be able to straighten out her legs, but moves with them underneath, from the knee down being flat on the ground. Can you tell me what is the matter, and a remedy? —W. A. B.

Answer—I am afraid your hen has rheumatism from liver trouble, brought on by overfeeding, with insufficient exercise, and I cannot hold out any hope of a cure at her age. If she is not feverish, she would be good for the table, but being very fat, and with this rheumatic tendency, she would never make a good layer, and the hatchet is the only cure for her. For the rest of the flock, give them Epsom salts in the drinking water for a week, and bicarbonate of soda for a second week; increase the amount of green food and meat, and cut in half the amount of grain, and let all of the grain be fed in the scratching pen to induce exercise.

RHEUMATISM IN THE FEET—I have a very fine Buff Leghorn rooster and he seems to have rheumatism in his feet. Do you know any cure?—Mrs. J. M. S.

Answer—Rheumatism may result from long exposure to cold and moisture; it may be produced by overfeeding of meat; induced through the underfeeding of vegetable food and is helped along by previous rheumatic tendencies of ancestors.

Treatment—Bathe the feet and shanks with the following: One cupful of vinegar, one of turpentine and a heaping teaspoonful of saltpeter, mix in a bottle and shake well before using. For internal treatment there is no better remedy than iodide of potassium. This is given in the drinking water, fifteen grains of iodide of potassium to every quart of water. Give in small dishes, so that it all may be used while fresh and thus avoid waste from having to throw away any because it is mixed with dirt. Common cooking soda, one level tea-

spoonful to each quart of water, or salicylic acid, one grain a day, has given good results, but the iodide is the best and most satisfactory. Give plenty of green food.

ROUP, BRONCHITIS, PNEUMONIA—(F. M. C., California)—Can you favor me with a little information which I fail to locate in your valuable book and it covers the ground very well. On a cold and windy night, two weeks ago, a careless boy left a window open in a house, allowing a strong draft to blow on my precious four-months-old pullets. Consequence, about half of them (586 all told) came down with bad colds. Some developed roupy catarrh, others eyes swelled close shut. Sprayed nostrils with glycothermoline and carbolic acid. No good effect noted. Put roup cure in drinking water and dipped head in same. Majority are improving. There is one phase of disease that puzzles me, and of course it attacks the largest and finest pullets. They seem to have difficulty in getting their breath. Act like a chick with the gaps. Open their mouths and gasp with a strained, worried look on their faces. Live about twelve hours and die choking to death in one last convulsion. These so affected have not so much odor at nostrils as majority. No mucous spots in throat. Throat seems to be full of phlegm. Don't eat at all. Spraying throat with glyco-thermoline and acid, and painting with iodine or running feather saturated with coal oil down windpipe offers no relief whatever. No one around me seems to know of any remedy. If you can diagnose it and suggest a remedy, will appreciate it greatly, as I hate to lose chickens when they get this old, and I put great faith in your suggestions.

Answer—I sympathize most sincerely with you in your trouble from your beautiful pullets taking cold, and wish I could help you. I think you have been doing all that was possible. You see, hens are very much like human beings. One person will have neuralgia from a draught, while another will have a sore throat, and while from the same cause one may have catarrh, in another the trouble will be bronchitis or even pneumonia. Now, I think with your pullets, some of them have catarrh, others swell heads, and with others the catarrh has gone down lower into the bronchial tubes and possibly into the lungs themselves.

Now as to treatment. If I remember rightly, the roup cure you are using is made principally of permanganate of potash and bluestone (sulphate of copper). Both of these are excellent germicides and by killing the germs of the catarrh or·roup, they prevent their multiplying, and give nature a chance to recuperate. I think, though, the roup cure is more effective than the severer medicines, such as turpentine and carbolic acid, so I now recommend that your roup cure be given in the drinking water, at the same time dipping the head in the same. Or you can put one cupful of kerosene oil into two parts of water. The oil will float on top; dip the fowl's head slowly under this, holding it there while you count three. It will sneeze and cough and you must wipe off the mucus with a rag and burn the rag.

With some of the fowls the catarrh will go deeper and for these I think the peroxide of hydrogen, spraying the throat well, is the best, giving always the permanganate of potash and bluestone in the drinking water.

For those that have developed bronchitis or where you think the bronchitis may be just commencing, give aconite, one drop in a teaspoonful of milk, twice or three times a day. The symptoms you describe are exactly those of bronchitis, so I feel confident in recommending the aconite. Dr. Woods recommends the "Aconite, Bryonia and Spongis mixture," but I have not tried it. The mixture is "ten drops of the tincture of each in an ounce of alcohol. Use a teaspoonful of this in a quart of drinking water." I think this might be very useful, especially at the commencement of a cold or bronchitis. Dr. Woods says that two doses will often effect a cure. Or you can get this in tablet form at. the drug store. The tablet (1-100 of a grain in strength) can be given one to each bird two or three times a day or twelve tablets in each pint of drinking water.

I have found a teaspoonful of honey with five drops of eucalyptus oil, twice a day, to be an excellent cure. The honey is very soothing and is also nourishing and sustaining. Bronchitis is a very debilitating illness and the fowl should be fed only liquid nourishment, such as raw egg beaten up with half the amount of milk, about two teaspoonsful every two or three hours. I have given a tablespoonful of milk or milk with honey mixed. I have a small "invalid drinking cup;" it is a narrow cup with a spout like a teapot, which I have found very useful and handy, as I could insert the spout a little ways down the throat of the hen and none of the liquid would be spilt. A child's toy teapot with a rather long spout will answer the purpose, but an invalid drinking cup, costing ten cents, is extremely useful and worth many times its price for chickens. You can use a dropping tube also for administering liquid medicine. I realize that with the large number of fowls that you have you want an easy and quick way of doctoring, and the only way is by the drinking water.

In cases of cold or the cold going deeper, as in to bronchitis or pneumonia, fowls need very easily digested, light and nourishing food. I have found nothing better than bread and milk. To this can be added a little bran, or a few eggs can be beaten up with the milk before putting in the bread if you think necessary. You did perfectly right to segregate the fowls. Colds of all kinds, even pneumonia, are infectious.

I would strongly advise you to house your hens in open front houses. In this way there would be no draughts from windows left open. Open front houses are a preventive of both bronchitis and pneumonia.

I have found that the pills or asafoetida and quinine which I recommend in my book, if given at the very outbreak of a cold, frequently cure with one dose; also the mixture, No. 5. This is Mr. Hunter's old remedy and has been found successful by hundreds of people.

ROUP—HOW TO CURE IT—I have over a hundred hens, all breeds. A good many of them are sick; I have tried everything, but to date I have not found anything to do them good. A yellow, hard substance that has a very bad odor forms in their mouths and eventually in their windpipes and they drop over dead. I have lost about thirty inside of a month. I feed chopped corn and wheat, with plenty of Pratt's chicken food. Use Conkey's Roup Cure and bluestone. They run at the nose and their eyes swell shut; others look fine, combs red, and you would not know anything was wrong with them until they fall over dead. Can you tell me what is the matter with them and what I am to do with them? I paid $1.00 apiece for my hens and it is hard to see them all die and not know what to do for them.—Mrs. R. B.

Answer—I am very sorry to say that it is diphtheritic roup that your hens have—very like diphtheria in children.

It is a germ disease. At first the hens take cold and the germ then seems to take root and the yellow leather-like spots commence to grow and continue until they choke the fowls.

The first thing to do is to separate the healthy fowls from those that are sick and disinfect the premises thoroughly. Discover if possible what is giving the fowls a cold. The usual causes of cold are a draught in the sleeping room, a narrow draught that strikes on the fowls as they roost, caused by a crack or a knot-hole, or a house that has no ventilation; too much crowding at night, which makes the fowls hot and sweaty, and they take cold when they come out in the morning fresh air, or roosting out side in the rain and dew. Lice will also give them cold and will carry infection from fowl to fowl. When one fowl has a cold, the others are very likely to catch it from the water, from the food or from contact in sleeping on the same perch. I explain this so you may decide for yourself what is causing the trouble and may use preventive measures and stop their taking cold.

Now for some cures: See page 105.

A bit of bluestone (sulphate of copper) as large as a navy bean, in a quart of water, is an excellent remedy and preventive. Blue stone is a germ killer and when it is in the water it will kill the germs that float off the chicken's nostrils, and that would infect another fowl. It also kills any germs that it may reach in the sick fowl's nostrils and so dries up the cold in the head. Of course it is a strong astringent poison and should not be given in stronger doses than I have indicated. Also keep those pretty bits of blue out of reach of the baby. Rub the heads of those that have watery eyes with carbolized vaseline and put a little into the nostrils and in the cleft of the mouth.

For those that have the white or yellow spots, spray the mouth or swab it with peroxide of hydrogen twice a day. Use it half and half water. The peroxide of hydrogen kills the diphtheria and will prevent its developing. There is a possibility that the spots may be canker in some cases (those that are apparently not very sick) in which case get four grains of sulpho-carbolate of zinc, dissolve in one ounce of distilled water and paint the spots lightly. This will kill the germ of canker. It is not the same germ as the diphtheria, and the two medicines cannot be mixed, as they may be said to neutralize each other. If you are not sure which disease it is, you might doctor one day with peroxide and the following day with the zinc.

Add to the diet of the fowls onions chopped finely, with a teaspoonful of cayenne pepper for a dozen hens, or if you can get them, grind up chili peppers and give a tablespoonful in the food or mixed with bran.

SCALY LEGS—Will you be so kind as to explain what kind of disease my hens have? I am a green man in the poultry business and bought the hens from several places, with the intention of having in the shortest time a sufficient number of egg producers. Among the purchased birds there were about sixty with scaly legs. I inclosed them in a separate yard, 30 x 40, fed them abundantly, and every morning they were urged to pass through a tray with coal oil. After ten days many of them had legs clean from scales, but some became weak and droopy. They walk with difficulty and keep their tails down. They grow worse every day. I killed two of them and found that about half their bodies were covered with yellow scales like a sort of bad skin which you can easily tear off. Is it a contagious disease, and what shall I do with the sick birds?—F. P.

Answer—Poor hens; it is not a disease. It is the coal oil that wets their feathers and that blisters the skin. Those that have been much wetted on the feathers with the oil are probably too badly burned to recover. The others will get well in time, but it will greatly delay their laying.

Do not try again such heroic treatment. It costs you too much. Next time mix one spoonful of lard with one spoonful of coal oil and one spoonful of powdered sulphur; rub the legs with that twice a week.

SCALY LEG.—Scaly leg does not appear without the irritation due to a parasitic insect. This parasite comes from another fowl, or possibly from an infected house or brooder, and works its way in between the scales of shanks or toes. Scaly leg passes from one diseased bird to another on the roost or is contracted by chicks when with the mother hen. A single case of scaly leg on the plant is a source of danger to every other bird.

Scaly leg is so easily cured that no intelligent poultryman is excusable for having its presence on his place for over a week. Every bird bought should be examined for scaly leg, and any doubtful one receive immediate attention. If you at any time find several cases on hand I would advise the applying of the proper treatment to every bird on the place. This is not much trouble and prevents the cropping out of new cases in a short time.

Paint the perches with lice killer or kerosene and naphthalene flakes.

Scaly leg comes from the scale mite and is very infectious.

SWELL HEAD—My chickens are dying off awfully. Many of them are good sized pullets. Their heads seem to swell and they go blind and just drop off. Some of them open their mouths and stretch and act as though something was choking them, but I cannot detect anything. They had mites, but have none now. We have a good yard for them, and an alfalfa patch and some shade trees. I feed them well, and am at a loss to understand. My neighbors on either side of us have the same trouble.—Mrs. F. K.

Answer—Your chickens have what is called "swell head" and roup. They have either caught it from taking cold or from the lice which they used to have, or by infection from the neighbors. I think probably there is a draught in their sleeping quarters, from a crack or a knot hole or it may be wrong ventilation. Stop these up and be sure the chickens do not live or sleep in a draught. Rub their heads with carbolated vaseline, and give each of those affected a quinine pill every other night for a week, and add a little poultry tonic to their food. I think as soon as you stop whatever may be the cause of their taking cold you will have no further trouble. Be sure to keep the sick fowls away from the balance of the flock.

EYES SWELLED SHUT AND WATER—Will you kindly tell me the cause of sore eyes? My chickens' eyes swell shut and water. I also have turkeys; their eyes swell underneath.—Mrs. C. J. N.

Answer—Your chickens and turkeys have lice and are taking cold. They are taking cold from either sleeping in a draught or sleeping in a place that is too close and hot, so they take cold when they come out in the morning. Remedy the cause and use one of the many roup cures, and also get rid of the lice. Lice go to the eyes to drink and so spread the disease.

TOE EATING—Can you tell me what causes little chicks to pick at each others toes? They will pick at one till the blood comes, then so many chase it that it dies. Then they start on another and sometimes they even eat the entrails out. I bought my chickens when they were a week old and fed them according to your directions. I first fed raw meat and cooked, then I tacked pieces on a board to keep them busy, but nothing seemed to stop them, and I took the one out with the sore toes. I gave lime and salts and charcoal. I hatched some dark colored chicks in my own incubator and with them I have not had any trouble in any way. I trust that you can help me.—H .L.

Answer—It is usually with the white or light colored chicks that we have this trouble. The little toes are so attractive and look so very good to eat that a lively chick will often try to taste his neighbor's toe and it tastes so good that he continues the performance and soon teaches the others. Dark toes are not so attractive looking, hence their immunity. You did quite right to add more meat and even a little salt to their diet, but the best way of preventing the trouble is to give the chicks chaff at least an inch deep in the nursery of their brooder. I have found that alfalfa hay or wheat hay cut in a clover cutter an inch in length make very good chaff for the chicks. I scatter the chick feed a little at a time, three times a day, in this, and the chicks scratch in it and find the grains and at the same time it conceals their toes from their hungry brothers. In this way you not only prevent this vice, but you make the chicks scratch many hours a day and that broadens their backs and develops the egg organs and strengthens their digestion, keeps them out of mischief, healthy, happy and busy. Try this plan and you will be surprised to find what extra fine layers you will have next year.

TUBERCULOSIS—A year ago I had the nicest Black Minorcas that anybody ever laid eyes on, but, alas, one after the other I had to kill. First they get lame on one foot, then their combs get very dark, almost black on the points; their appetite is poor and they get as light as a feather, and when I cut them open their liver almost fills up their whole insides, and the whole liver is thoroughly sprinkled with little white kernels; sometimes as big as a good sized head of a pin, sometimes as large as five cents, and I attend to them so good. Now, can you tell me what disease it is and how to prevent it after this? I feed lots of green stuff, milk, meat, wheat, barley and occasionally a mash of lots of carrots.—Mrs. M. R.

Answer—I am sorry to say your Minorcas have chicken tuberculosis. You gave an accurate description of the disease, and I am very sorry to have to tell you that there is no cure for it when once it has commenced. You may be able to prevent the young ones catching it by moving them on to fresh ground, and thoroughly disinfecting the yards and coops. Send a postal to the Experiment Station, University of California, for the bulletin on "chicken tuberculosis;" it is free.

TUMOR AND DROPSY—I had a White Leghorn hen die a week ago from an ailment which puzzles me. Have looked through what poultry books I have, but can find nothing touching it. The hen was swollen between the legs to an unusual size and got so bad it could not walk. Finally it died, and, upon opening it, at least a quart of water came away. The intestines were joined together in one solid piece. Can you tell me the cause and cure, as I have a Hamburg hen developing the same symptoms and would like to save it if possible.—J. L. W.

Answer—Your hen died of dropsy, combined with a tumor, probably ovarian. There is no known cure for this, as by the time it becomes visible, the disease has progressed too far, and is usually only discovered after death. Some hens seem more subject to this complaint than others, and I would advise you to get in fresh blood and keep the hens healthy by feeding an abundance of green food. The cause is obscure.

VENT GLEET—One of my hens and fine, large cockerel have a sort of diarrhoea with a very bad smell to it. It seems to scald the vent, which is red and swollen and there are scabs on it. Can you tell me the cause and cure of this?—Mrs. J. F. Y.

Answer—Your hen and probably the cockerel also have vent gleet. This is usually caused by an egg being broken inside the hen, which causes inflammation. It is, I am sorry to say, contagious, and the birds should be at once isolated and treated. Prepare a warm bath of water as hot as can be borne on the wrists, in which has been dissolved a tablespoonful of bicarbonate of soda to two quarts of water. Immerse the fowl's abdomen and vent in this hot water and hold the bird there from fifteen to twenty minutes. Then dry the parts with a clean cloth and give an injection of an infusion of green tea with five grains each of sugar of lead and sulphate of zinc to each ounce of the infusion, two tablespoonsful being one ounce. The sores and ulcers around the vent should be kept dusted with iodoform or aristol. Repeat the treatment once a day until the bird is cured. A dose of thirty grains of Epsom salts will help cool the blood. Feed lightly and give plenty of green food. If not well after two or three weeks, kill the bird, as the disease is not quite free from danger, for if the operator should touch his eyes accidentally before cleaning his hands, the result might be a most violent inflammation, and the disease is extremely contagious among the hens. One cockerel may infect all the hens.

WHITE COMB—My fine Orpington rooster is developing a peculiar disease. A few months ago he was in the pink of perfection, but his comb has become all covered with white spots, as though he had dandruff, and it spoils his appearance. I feed your well proportioned mash, wheat, alfalfa, crushed green bone, lettuce and cabbage; a mash every morning and corn or wheat for the evening meal. He is vigorous and active, the only trouble being with his comb. If you will kindly tell me how to treat him for this trouble, it will be highly appreciated.—E. R. T.

Answer—Your rooster has what is called "White comb." It usually comes from close air in the hennery and a total absence of all green food. It is a contagious disease and may be imparted from bird to bird, probably also from mice, rats, cats and dogs to birds. Young birds appear to be more susceptible to this disease than old ones. Put carbolated vaseline on the comb, and in the drinking water use twelve tablets of nux vomica and sulphur comp. 2X to each pint of water. Continue the treatment until cured.

———

WIND IN CROP—Will you please tell me the cause and remedy of my little chicks, from three to four weeks old, having a gas gather in their crops? When the crop is pressed, wind comes from their mouth and they stand around and gasp, but otherwise do not look droopy. They eat well, but in three or four days die. I lost quite a number last spring, almost every case being fatal. I have a hen with young ones and I would like to raise them without this trouble.—B. C.

Answer—The wind in the crop comes from indigestion. Indigestion comes from lice, colds, dirty water, and chief of all from wet mashes or from wrongly balanced food, and lack of hard, sharp grit to grind the food. I do not think the chicks with the hen, if she is allowed free range, will get it, but if there are any symptoms of it, put some lime water into the drinking water and give them pounded up charcoal. Give them also sweet skim milk to drink, as well as water and plenty of nice, crisp lettuce to eat. I am sure if you keep them quite clean, feed clean dry chick feed with plenty of green lettuce, grass or clover, cut up fine, you will not have any wind on the stomach with your chicks. A little bicarbonate of soda in the drinking water will sometimes help, but prevention is the best cure.

———

LICE, MITES, TICKS AND WORMS

BODY LICE—I have about 100 White Leghorn chickens and I find that they have a large body louse, large yellow ones; what can I do to get rid of them? I think they are keeping my chickens from laying as they should.—Mrs. B. W.

Answer—Paint the bottom of a box or barrel with a good lice killer; put a little straw in to keep the paint from the feathers, then put the chickens in and cover them three hours. Then examine the hens and pull out all the feathers that have nits (lice eggs) on them, putting the feathers into a little can of coal oil. Then dust the hens with a good insecticide once a week or until you are sure all the lice are dead. Be careful to give the hens a spot of ground, well spaded up, mellow and a little damp. They will bathe in this and usually keep themselves clean.

———

DIPPING HENS—Would you be so kind as to let me know about dipping hens, etc? I have a flock of some five or six hundred. I notice some of them have lice and bunches of nits on their feathers. Whenever I have caught a hen I have greased her well, but this would take too long to go through the bunch. Is there any dip that would be strong enough, and do no harm to the birds, that would kill the nits with one dipping?—W. L.

Answer—Lice are supposed to hatch out the nits every five days, and when but a few days old commence to lay again and so keep on breeding indefinitely. Dr. Salmon says it has been estimated that the second generation from a single louse may number 2500 individuals, and the third generation may reach the enormous sum of 125,000, and all of these may be produced in the course of eight weeks. I do not know of any dip that will kill the nits with one dipping. Dr. Salmon recommends a dip of one per cent carbolic acid solution, or using creolin, as it is equally efficacious in killing insects and is less poison to the birds. It is used in the strength of two and a half ounces mixed with a gallon of water. I have used very successfully in the summer time when the weather is warm, the kerosene emulsion made as follows: Dissolve one bar of soap or one pound of soap powder in a gallon of boiling water; add to it a gallon of coal oil, churn for twenty minutes or until you wish to use it. Take one quart of this top solution and add it to nine quarts of water. Dip the hens into this, being careful not to allow any of it to go into their eyes or mouth, but thoroughly wet every feather to the skin.

This will kill every living louse, and if repeated in about five days will probably kill those that are hatched out in the meantime and prevent their laying any more nits. Tobacco water has also been strongly recommended as a dip, and chloro-naphtholium used as directed on the bottle.

THE SAND FLEA—How can I rid my chickens from a small insect known here as the sand flea? I have tried coal oil mixed with lard without effect. The hens scratch their heads so they become sore and some have died; others have had to be killed.—Mrs. F. A. F.

Answer—Those fleas are very hard to get rid of. Spray the henneries well with either the kerosene emulsion or good hot salt water, and while the ground is still wet, scatter on it air-slacked lime. Those hens that have sore heads should have carbolated salve put on them, after swabbing them off with corrosive sublimate. This will kill the fleas and cure the sores. Be careful not to let any of the corrosive sublimate get into the eyes or mouth of the fowls.

STICK TIGHT FLEAS—We have noticed a tick or louse on a few of our chickens and have discovered some of the insects on the perches. They resemble small black beads and are firmly imbedded in the skin. On some of the fowls we have used for the table we noticed a few red blotches on the skin. We would like to know how to get rid of the insects, particularly how to get them out of the hen house.—An Inquirer.

Answer—You have the stick tight fleas in your hennery. They are very hard to get rid of, being in some places a perfect pest. A friend of mine lost 500 out of 700 chickens last fall from this. I told him to spray very thoroughly with salt and water and he purchased 600 lbs. of salt, scattered it all over the hennery and yards and then turned a hose on them for several days in succession. He tells me now there is not a stick tight flea on the place. I advised him to get some corrosive sublimate diluted with alcohol at the drug store, take an old tooth brush and carefully apply with it the corrosive sublimate on any fleas he might see on the chickens, being careful not to allow any of the solution to get into the chickens' eyes (it would blind them) or into their mouths, as it is very poisonous. You can paint the perches with this; it will kill everything it touches. A saturated solution of salt and vinegar applied to the fleas on the chickens' heads or bodies will drive them away or kill them.

HEAD LICE—This time I write in desperation, hoping you may be able to give me a remedy. It is head lice I am fighting, and after working for almost five months, I am as far off from being rid of them as at first. I have done everything that I have ever heard of. I still find they have head lice and red mites besides. I hope no other beginner has had the trials I have had.—Mrs. W. F. K.

Answer—The red mites live in the houses or coops, except when they are feeding off the chickens, usually at night. The cure for them is to spray the coops thoroughly and constantly. You can keep them out of the coops by spraying once every three weeks, but if they once get in, you will have to spray twice a week until you get entirely rid of them, then once every three weeks, to keep rid of them. The head lice live on the heads of the chickens. They lay two or three white silvery nits (eggs) at the root of the feather. The eggs hatch in about five days after they are laid by the lice, consequently to completely destroy them, you should treat the chickens that have them at least once a week. The best way I know of is to take an old tooth brush, a bowl with nice hot soap suds in it and a few drops of the best carbolic acid; brush the chicken's head with this, being sure to touch all the lice and mites. This I know, is an excellent remedy, for I have tried it. Another given by a friend of mine is, get the druggist to mix some corrosive sublimate with the best pure alcohol, take the tooth brush and brush the chickens' heads with this, being careful not to let any of this get into the eyes (or it will blind them) or into the mouth, as it is very poisonous. This will not only kill the head lice and their nits, but it will also kill stick tight fleas, ticks and any insects. It is very difficult when once the pests get into henneries or on chickens to get rid of them. It is far easier to keep the enemy out by constant and thorough cleaning at frequent intervals, especially in the summer time. I find using tobacco stems for making the nests of setting hens a good preventive; besides this, I see that all the fowls have good dust baths in damp and mellow earth.

MITES—We are fighting mites, but apparently with no success. We hired a man who makes poultry ranch spraying a business. We paid him $10 and he guaranteed to rid the place of the pests, but they are worse than ever. He uses lime, sulphur and carbolic acid. Is there any way corrosive sublimate could be used as a spray, and would it be safe for the hens in the houses? How long would the hens need to be kept out after the spraying was done? Am having the worst possible luck with my chickens. Have probably hatched 550 chickens this year and have less than 200 now. When a week to ten days old they begin to droop, refuse to eat and starve to death. What is the matter? No bowel trouble; no cold; no lice, or only a few. Does cholera ever attack such young chickens, and if cholera, would they not have bowel trouble? Would greatly appreciate an immediate answer, as the mites get all over me and drive me nearly frantic.—Perplexed.

Answer—The thing that is killing your little chickens is not cholera, otherwise they would have bowel trouble; it is only the swarms of mites. If they drive you nearly frantic, think how the chicks must suffer. The mites simply drain the life out of them. The corrosive sublimate can be put on with a spray, but it is dangerous to do so, as if it splatters into the person's eyes who is spraying, it may blind him for life. One pound of this costs $1.25 and that is sufficient to make 120 gallons of the solution. As it takes some time to dissolve in water, it is usual to dissolve it in alcohol. I have used it dissolved in alcohol to paint henneries and nest boxes, and it will destroy all insect life. You must turn the hens out of your henneries for several hours, or until the walls are dry.

INSECT POWDER—Mrs. C. B. F., Los Gatos—I do not think the "flea powder" you mention would kill the little turkeys, but as you ask what I use, I will tell you. It is here called "Buhach," and can be bought at any of the poultry supply houses. It is made from the "Pyrethrum" daisy and is perfectly harmless to all fowls, from tiny canaries to mammoth turkeys, but deadly to insects. It contains a small quantity of an essential oil which asphyxiates all insects, fleas, ants, lice, etc. It must be kept in an airtight jar or tin box, as the essential oil easily evaporates. Next in value come the insect powders, the foundation of which is tobacco dust.

The kind of lice that are so deadly to little turkeys are the same as the head lice of chickens. They are to be found on the heads and necks of the turkeys, and also on the large feathers at the edge of the wing. They seem to sap the life out of the turkeys. I always rub the "Buhach" powder well into the down on the head and at the roots of the wing feathers, whether they have signs of lice or not, for it is better to be sure than sorry.

TICKS—In trouble again. We are renting a place until we can build on our own, and every building on it is simply alive with little brown ticks; they bury themselves in the heads of the chickens, the ears of the dogs, the feet of the animals and all over our bodies. What shall I do? Please tell me and tell me quick. A neighbor says lard and carbolic acid on their heads and spray with distillate, but neither seems to do any good so far. I am out of the chicken business since moving here, except a few for our own use. Yours sincerely. —J. J. W.

Answer—The easiest way to get rid of them is to pour coal oil over the buildings and then set fire to them, but as you are in a rented place, that would scarcely be possible. The next best plan is to paint the place thoroughly with corrosive sublimate; it is what I recommended to you for the plague mites at your other place. Ticks are one of the worst plagues in Southern California. They are so thin and flat that they hide between the singles and boards. They really are not thicker than a bit of paper, and nothing kills them but the corrosive sublimate (bi-chloride of mercury). This can either be put on with a brush or be sprayed on the houses. You remember that it is very poisonous, and great care must be used in handling it. When once your coops are free of ticks, or other vermin, you can keep them so by spraying with kerosene emulsion that I have so often given. Distillate, liquid lice killer, coal tar and other preparations of carbolic acid or creosote are all good to keep out vermin, but I know they will not drive out ticks.

DEPLUMING MITES—Two years ago I started to raise White Leghorns, commencing with two cocks and twelve pullets of as good strain as I could secure at the time. This spring I had a splendid looking flock of 100 females and twelve males. They were beauties, but recently developed the feather-pulling habit and are now a sight. Never in moulting time have I seen poultry look worse. Many of the hens look as though plucked for market, and not one of the roosters has a vestige of tail. The hens still keep up laying as well as before (from fifty to sixty-five daily), but I cannot believe this will hold out in their present condition.

I have them on a two-acre range and feed them cut green bone in large quantities four times a week in addition to all the other grains obtainable. My experience can only suggest two causes for such a state of affairs: 1, Insufficient animal food. 2, Close confinement. But neither of these causes enters into the present state of affairs. Can you advance a reason and suggest a remedy? By so doing you will greatly oblige one who is getting interested in raising fine looking birds.—F. S. S., Tucson, Ariz.

Answer—Your birds have what is called "Depluming mites." The principal symptom of this trouble is a loss of feathers from spots of various sizes, situated on different parts of the body. The feathers break off at the surface of the skin, and at the root of the feather is seen a small mass of epidermic scales which is easily crushed into powder. A microscopic examination of this powder reveals numerous mites and the debris which they produce.

The disease appears in poultry yards as a consequence of the introduction of one or more birds already affected. It is readily communicated, develops rapidly and in a few days a whole flock is contaminated. It usually begins on the rump and spreads rapidly to the back, the thighs and the belly. An infested cock will rapidly infest all the fowls in a poultry yard. Often the head and the upper surface of the neck are affected early in the course of the disease. The feathers fall off at all these points and finally the skin is denuded over a a large extent of surface. The large feathers of the tail and wings and the wing coverts are generally retained.

The denuded skin presents a normal appearance; it is smooth and soft, of a pinkish color and not perceptibly thickened. By pulling out the feathers which remain near the invaded parts, it is easy to find, with fowls, a mass of epidermic scales at the end of the quill, which contains a number of parasites. The general health of the birds is apparently not disturbed. They remain in good flesh and continue to lay as though they were not affected. It seems probable that much of the irregular moulting, feather pulling and feather eating are due to the irritation caused by the Sacroptes Laevis.

The treatment for this is not very difficult, but must be persisted in until a cure is effected. Carbolic salve should be rubbed over the affected portions of the skin and the adjacent parts, or a salve may be made by mixing one part of carbolic salve, one part of flour of sulphur, one part of powdered aloes with ten parts of lard or vaseline.

A large surface of the body should not be covered with strong carbolic acid preparations, on account of the danger of absorption and poisoning. The affected parts of the body may be rubbed every fourth day until a cure is effected. It is well to finish the treatment by dipping the birds in a two per cent creoline bath and to whitewash the houses with carbolated whitewash. This will kill any mites which may be left in the feathers or about the roosts.

Worms From Wild Birds — Some years ago my fowls became afflicted with a round worm, also tape worms, and in one article you mentioned several remedies, such as santoine, turpentine and tincture of male fern. I dug up the yards and seeded to green feed, but all to no purpose; it has practically driven me out of business. Last spring I invested in some outside stock (just hatched baby chicks), but they also became infested, although they were on new land. However, I managed to keep down those pests by occasionally dosing the hens with the above mentioned medicines. We do not feed anything unclean to our fowls and it always has been a puzzle to me where such worms came from.

A few days ago our house cat brought home a small bird, which she began to devour on the house porch, but leaving the intestines, out of which crawled two good sized round worms such as fowls have. As we live in the woods, do you think this has anything to do with it? I am almost afraid to start my incubators this season, as it may only result in future failure.—W. E. B.

Answer—Your fowls undoubtedly get the worms as the wild birds do, from the droppings or eggs of worms from the other birds. By the persistent use of turpentine, using 60 drops in a quart of water, or mixing it in that proportion in the food, for a week at a time, you can get rid of them. Also disinfect the ground. The only thing that I can see is for you to keep up this treatment, for a week every two months, giving turpentine either in the food or water. I would not be discouraged because that is a sure remedy and by watching and noticing the droppings, you need not fail in rearing the chickens.

WORMS FROM PIGEONS—My chickens' gizzards are affected by red worms about the size of a pin. All the stock I raised last year seemed affected, although the eggs came from different places. I have the Brown Leghorns, Brahmas and R. I. Reds. I feed all the various grains, plenty of greens and good meat and bone. The only thing you recommend that I have not fed is charcoal, still as chicks they got it in the chick feed. I have given them turpentine in food and water at various times and it seemed to have the desired result, but today I learned different, the gizzard is penetrated and has a sore spot caused by these worms. All the stock in different yards are affected.

I get plenty of eggs and the chickens look good, combs nice and red, nevertheless I find them all affected the same way.—Mrs. G. S. L.

Answer—I have been through the same trouble myself and so can help you. I found out that my chickens were getting the worms or the eggs of the worms from neighboring pigeons. The droppings of the pigeons contained the eggs of the worms and in a short time the droppings of the chickens also had them and the other chickens ate them, and so on they kept increasing. First of all I gave the chickens the turpentine which I recommended to you. A teaspoonful in a quart of water. Mix the food with that water, also put a teaspoonful in a quart of the drinking water and allow no other water for drinking. Keep this treatment up for a week. Meanwhile clean up the yards by having them either ploughed under or dug up and a crop of some kind planted, something that will grow quickly, such as wheat or barley, and as far as possible destroy the birds that are bringing you the trouble, for I cannot but think it must be pigeons or some other wild birds. The

worms will kill the young chickens, but they do not always kill the older fowls. Sometimes the worms come from unclean or spoiled food, from "webby" grains and bad animal food. You will have to discover for yourself where they are getting the worms from and cut off the source of supply.

INTESTINAL WORMS—I wish a little information and advice in regard to a valuable Buff Orpington cockerel I own. He has become mopy and goes away under the trees by himself, and has lost over half of his weight in a month. He eats like a horse, though, of everything I give my hens, but shakes his head an awful lot, as though something was wrong. I looked in his throat and it looks all right. He has changed in color from a light buff to a very dark red since acting unwell, and has grown to be a homely, dopey bird, from a real beautiful lively one a short time ago.— M. J. Q.

Answer—I think your Buff Orpington cockerel has intestinal worms. You had better give him 25 drops of spirits of turpentine on a lump of bread, or in a spoonful of water, and follow that immediately with two teaspoonfuls of castor oil. Keep him shut up so you can watch the droppings and remove and burn or bury them deeply. If you do not find worms in his droppings, give him ten drops of tincture of male fern on a lump of sugar, followed in an hour by a dose of castor oil. This is for tape worms. Both the remedies should be given after twelve hours or more fasting.

Dr. Sanborn says: If you suspect worms, try to remove them. Dissolve in the water that is to be used for mixing the mash, two grains of santonine for each bird to be treated. Mix a small amount of mash, quite dry and add castor oil. one-half teaspoonful for each bird. Feed this to the suspected birds, watching for the results of the "worm treatment." All droppings should be collected often and put out of reach of the birds.

SEVERAL KINDS—I am in despair and it is lice, lice, lice. We have Brown Leghorns, and as they will not sit, we borrowed a setting hen and she only stayed with us long enough to give our hens a supply of grey head lice. When we discovered them we went to work with a lice killer, sprayed the coops, ground and nests, put the chickens in a box and left them three hours. We also used crude oil, poured gallons on the ground, painted nests, roosts, etc., but still the lice stayed on the hens' heads. Last week we bought six Buff Orpingtons; yesterday we found they were alive with body lice, yellow lice, especially around the vent; there were thousands; then we examined the Leghorns, found they were infected also. What shall we do? Do you think it would hurt them to wash them now with the kerosene emulsion? Am afraid it might give them a cold.—Mrs. C. S. B.

Answer—What I should do were I in your place would be to get some Buhach powder, rub it well into the chickens' heads for the head lice, and well into the fluff under the wings and on the backs for the body lice, then put the hens, six or a dozen at a time, into a large size dry goods box, at the bottom of which is a newspaper thoroughly painted with a good lice killer; cover the top of the box with a carpet and leave them in for three hours, then look them over thoroughly and pull out every feather that has nits on it. The nits hatch out about every five days, so in a week's time look the hens over again, powder them again, and again put them into the box painted with the lice killer. Two applications should cure them. After this, once a month, at night, powder them with buhach and look them over occasionally, and, if necessary, go through the performance again. You can paint the roosts with lice killer, but do not put any in the nests, for it will not only flavor the eggs, but will kill the germs and make the eggs unhatchable. The best thing to use for the

nests is a kettleful of boiling water with a large handful of salt added to it, or scalding soapsuds, putting in fresh straw, or, better still, making the nests of tobacco stems. You can get these for 25 cents a gunny sack full.

———

SPRAY FOR HOUSES—Last summer I found a recipe in one of your articles for spraying hen houses. I used it to good advantage, but have misplaced the recipe and cannot remember the mixture exactly. It was composed of coal oil, carbolic acid and soap, with a certain proportion of water. If you will kindly send it to me, I will appreciate it.— C. W.

Answer—I gladly send you the recipe, which is excellent. I have used it for ten years or more. It will kill fleas, lice, mites or any insect pests in the henneries. It will also thoroughly disinfect the premises from infectious diseases.

Dissolve one pound of hard soap (or soap powder) in one gallon of boiling water, remove from the fire and add immediately one gallon of kerosene and one pint of crude carbolic acid. Churn or agitate violently for twenty minutes or until you want to use it. If the oil and water separate on standing, then the soap was not caustic enough. Add to this ten gallons of water.

I keep the stock solution on hand, dip out a quart and add to it ten quarts of water and use it for spraying the houses once every three weeks in summer and every month in winter. Putting it on hot in summer and slopping it well into dark and dusty corners will kill fleas, which are exceedingly troublesome on sandy soil in this part of the country.

———

FEEDING IN GENERAL

FEEDING SYSTEM—I am not perfectly satisfied with my feeding system and I follow yours on the food question. I note that you advise dried blood and other food dried in the oven, green cut bone and bone meal. Would you advise boiled liver, lungs and scraps instead of prepared meat scraps? Are ground clam shells good in place of cut bone? Could there be any danger from feeding too much ground shell? Should gravel be furnished to chickens to pick from?— D. F.

Answer—Boiled liver and lungs chopped fine are excellent for fowls. I prefer them to prepared meat scraps. They must be fed while fresh, as spoiled meat may poison the fowls. Clam shells cannot take the place of cut bone. Crushed oyster and clam shells contain lime, which is very good for making egg shell. There is no danger of the hens eating too much of this. Gravel or grit should always be furnished to chickens.

ANIMAL FOOD FOR FOWLS—Kindly inform me as to the difference, if any, between beef scraps, beef meal, meat meal and blood meal. Which is considered the best to feed laying hens and growing chickens? I have fed beef scraps for nearly a year and had good results from it; at least I think I have. If some of the others are better, I would like to know what one it is.—G. K. W.

Answer—Beef scraps, beef meal and meat meal are the same, only the latter is ground finer than the former. Blood meal is made from the blood, cooked, dried and ground. Pure dried blood contains more protein than the others, therefore is considered better in most cases. The beef scraps and beef meal are the refuse of the slaughter houses, heads, lights, etc., boiled down or cooked with steam, pressed, dried and ground, and are frequently called tankage.

If you have a good brand, keep to it, because some are no good, and if allowed to become damp or heated are injurious to the chickens.

BAD MEAT—I had twelve laying hens, they averaged seven eggs a day, were healthy and never were sick until I bought five cents' worth of green ground bone from a wagon that passes my door. It was wet and slimy, and smelled, but he said it was all right. I gave it to the chickens at noon; fed them nothing else then. At four o'clock I went out and found two dying and six more droopy and by eight that night had lost eight. Next day two large Buff Orpington hens died. I looked for some of your remedies giving asafoetida pills and the soda you spoke of in the water. I showed the bones to the butcher, and he said he never heard of such a thing as spoiled meat poisoning chickens. He sold it when it smelled like that all the time.—Mrs. D. M.

Answer—That meat poisoned your chickens evidently. It is called ptomaine poisoning. Butchers sometimes put formaline or some preservative on the meat, which has a very poisonous effect on chickens, but yours were undoubtedly poisoned by the putrid meat. You had better not buy any ground bone unless it is quite fresh.

BLOOD MEAL—Will you please tell me how much blood meal to put into the mash for thirteen hens, or in other words, what proportion for each hen?—L. S.

Answer—Half an ounce per hen every day at this spring season of the year is about what they need of blood meal mixed in the mash. Weigh out enough for the thirteen hens and measure that in a cup or by a spoon, then you will know how much by measure.

BEEF SCRAP—Is beef scrap, sold at the poultry supply houses, good for fowls? What is it and how much should be given per hen?—J. F. Y.

Answer—Beef scrap is excellent for fowls when it is good. It is made from refuse of the slaughter houses, heads, lungs, liver, etc. It can readily be detected if unfit for food by pouring boiling water upon some and if the odor smells of decayed meats it is unfit for use. Some put it in dry mashes, others put it in hoppers and allow the fowls to eat of it as they like, either way is good.

Examine also by placing a small quantity upon a piece of white paper and noticing carefully pieces which look more like dark brown glass; these are hoof and horn, very rich in nitrogen but cannot be digested by fowls.

Beef scrap, if kept in a warm or damp place, sometimes become lumpy. If you break open the lumps white threads may be seen in this. This is a very poisonous fungoid growth and will poison the fowls. If rats or mice are allowed to run over the beef scraps their droppings will also moisten the meal and render it poisonous. Thousands of chickens are lost each year by these poisons.

Always examine beef scrap before buying any great quantities, and reject any that has fiber or hair, hoof and horn, as it is unfit for food.

BEET TOPS—Will you kindly tell me if beet tops are a good green food for ducks? Also for fowls and turkeys? Are they as nourishing as alfalfa? My hens are not laying well. The eggs have suddenly dropped off, and I did not know but what the cause might be beet tops.—J. S .Y.

Answer—In September one is glad to get anything green for the fowls, ducks, geese or turkeys, to eat. Almost anything green is better than nothing, but alfalfa contains more protein than any other green food except white clover. The per cent of protein in white clover is 15.7, and in alfalfa 14.30, while in beet tops it is only 1.3. By this you will see that alfalfa is worth about 14 times as much as beet tops. There is about as much protein in alfalfa as in wheat bran. You complain that your hens do not lay. I think probably they are moulting. You cannot expect hens to lay all the time without taking a rest.

DRY HOPPER METHODS—I write you regarding the dry hopper method of feeding. How much space do you leave at the bottom for the feed to come through, and how wide do you leave the space for the chickens to eat out of? We made one, but it is not a success, for the box is bloody from their combs hitting against it. They stand and eat all the time and do not go and drink as you say yours do.—D. S. M.

Answer—I had the same experience with hoppers injuring the combs of the fowls, and now I make my hoppers like those used at the Maine Experiment Station, simply a box with a roof over it. The box is twenty-four inches long and eleven inches wide. The sides are cut like a gable, the highest point being sixteen inches high. The gable roof keeps the food dry and the hens waste scarcely any of it. The roof lifts off, or can be slid back to fill it.

DRY MASH—Will you kindly inform me as to the best method of feeding calfalfa meal to hens and pullets? I use hopper constantly filled with dry mash consisting of bran, shorts, feed meal and beef scraps, accessible at all times, and would much prefer adding the alfalfa to this. Or would you advise soaking it in water and feeding it separately? The fowls get grain twice a day and now if I add the alfalfa to the mash what proportion shall I make it? Also, is it as well to add the charcoal, two or three per cent, to the mash or feed separately. I wish to simplify the routine work as much as possible. —Mrs. O. K.

Answer—I advocate adding the calfalfa meal to the dry mash . It is the same as alfalfa meal. It would make a very good ration to simply add one part of calfalfa meal to your present mash, making it one part each of bran. shorts, feed meal, beef scraps and calfalfa meal. I feed this with excellent results, but at first the hens did not like the calfalfa, so I only added one iron spoonful, increasing the dose every day, adding one more spoonful until, within a month, they were having the right proportion. You can mix the charcoal in the same way, but I prefer to keep it separate with the grit and the crushed shell.

EXERCISE FOR FOWLS—I was greatly interested in an article of yours on feeding. You say give a hen a chance to work and no matter how fat, etc. Now what interests me most to know is just how you manage to give them plenty of work in a limited space. We, who occupy only a village lot, will be greatly helped if you will tell us how to keep the hens busy in such limited quarters. —G. P. C.

Answer—To keep hens busy, give them what is called a "scratching pen." Put a 12-inch board across one corner of your lot and fill that full of good wheat straw or hay; scatter all the grain you feed in that, and the hens will work all day digging out the grain; every grain they scratch out they will bury two, and so will keep up the exercise. If you are feeding the hopper method, put the hopper at one end of the pen and the water vessel at the other end; this will give them the exercise of walking back and forth. You can also hang up a cabbage for them to jump at, but scratching is the natural and best exercise for developing the egg organs.

TOMATOES—Do tomatoes tend to make the hens quit laying?—J. W.

Answer—Tomatoes will not do the hens any harm unless fed in very large quantities. There is not much nourishment to them and consequently they will not improve the laying qualities; otherwise a reasonable amount will benefit the hens.

FOR YOUNG AND OLD STOCK—I am very much interested in your articles and would like to ask you for a little advice. Being away from home all day, I have to feed in the morning enough to do all day. This I can manage for the old stock by feeding scratch food in the litter and dry mash in hoppers. But how can I manage the growing stock? Please give a formula for dry feed. Do you consider the scratch food sold by the poultry houses good food for the young stock? My chicks will not eat the baby chick food after a week or ten days. I also give them lawn clippings or lettuce every evening.

Is a handful of scratch feed to the hen once a day enough where they have the dry mash and table scraps? Is cracked corn good food to feed alone to young stock? I have Rhode Island Reds.—R. L. P.

Answer—Your questions relate principally to the feeding of the young stock, and you do not say whether you want to keep them for fattening for the table or for future egg layers. There is of course a difference in the way of feeding, or rather in the quality of the food to be given to them. However, I will tell you the way I feed for egg laying. As soon as I think the little chicks will eat whole wheat, I add it to the baby chick feed, a small quantity. If they pick it up quickly I add more each day, and in a few days I give also some kaffir corn or finely cracked corn. It should be finely cracked, as it is difficult of digestion. When it is too long in digesting, the corn ferments in the gizzard and that gives the chick diarrhoea, which often proves fatal. We never want to overtax the digestion of a chick, so I give corn carefully. This applies to the last question in your letter —it is not good to feed corn alone. It has been clearly proven that chicks do better, grow more quickly and mature earlier if they can have a great variety of seeds to eat. This is the reason we prefer to buy the chick feed already mixed from the supply houses. They have greater facilities for getting a variety of grains than we have.

When the young stock is old enough to eat the wheat and kaffir corn, they can be fed as you do the old hens, only remember to give them nice, clean litter to scratch in. It will need renewing oftener than that of the old hens, for if it gets foul and they pick up some of their own droppings, you will soon have a set of sick chickens. Feed the grains in the scratching pen to the little chicks, and also give them in a hopper of bran, alfalfa meal, corn meal, ground bone and either granulated milk or dried blood in equal proportions. The little chicks will prefer the grains in the scratching pen and eat those the first, which is just what they want, but if they are hungry they will go to the hopper. Most of the poultry supply houses now

make an excellent scratch feed; they realize the need of it and are able to mix it scientifically. I always buy from them, and if I think there is too much corn and that my fowls will become too fat, I say, "Please economize on the corn." You will find most of the poultry supply houses willing to mix the scratch food just as you want it. You are feeding the mature stock all right. One handful of the scratch feed in the litter is about right for the hens. The green food is quite important, the lawn clippings should be of clover or as much clover as possible, for the blue grass becomes so hard and stiff as the summer continues that there is not much nourishment in it and the hens will not eat it. Lettuce is good, but sometimes quite expensive and difficult to get, but there is another green food that has been found excellent and is within the reach of any one. This is sprouted oats. Take half a bucket of oats, pour warm water on them and leave them covered all night, then spread them in boxes. Any box will do. Have the oats about two inches deep and keep them damp. In four or five days there will be a mass of tender green sprouts. The hens will eat eagerly of this. A friend of mine has also done this with barley for many years with great success. This green food is as good for the young stock as for the old.

In your place I would feed as you do, throwing scratch food (a handful to each fowl) in the litter in the early morning, keeping the dry mash in the hopper, and feed the green food in the evening. Some of it may be left till morning, but will not wilt much, and they will eat it the first thing. Be sure they have plenty of water and have it shaded from the sun, either in a box on its side or in some sort of shelter.

MIXING FOODS—I want to ask you if there is any good reason for not mixing foods at the same meal. Prof. Jaffa of the U. C. said on one occasion that it was best not to mix foods—in feeding wheat, to feed that alone; the same of barley or of corn. Make either an entire meal. I have observed in feeding my chickens that they seem to enjoy a variety of grains fed together. Which method would you think best?

I am feeding rolled barley dry. Would you think it better to soak it? I give the mash at noon, dry, and green feed morning and evening. The fowls seem to like the green feed better at those time than at noon.

Answer—The reason Professor Jaffa thinks it best not to mix foods is because some hens will pick out all of a certain grain in a greedy manner, and by giving only one grain at a time, they are forced to eat what he chooses to give them. I would not venture to differ from so learned a man, but, like you, I notice my hens enjoy a variety, so I give it to them, and for the little chicks I am positive a great variety is by far the best for them. I found that the hens enjoyed an occasional feed of soaked barley, so I poured scalding water over a few pailsful of barley, covering it with gunny sacks to keep in the steam and when thoroughly soaked, fed it to the hens.

———

How Much to Feed—Can you tell me how much feed an average Leghorn should have in weight with a free range of two acres of alfalfa? Is green ground bone necessary all the year round or only in the winter? My hens will not lay and I may not be feeding right, although a few Wyandottes I have are too fat, but they get exactly the same food as the Leghorns. I have 72 hens and only got 12 eggs yesterday. Am not satisfied with the results and desire to have them do better.

Answer—An average Leghorn hen should have in weight for every pound weight of hen an ounce of food. As Leghorns weigh about four pounds each, they would require about four ounces of food each per day. Animal food of some kind is necessary for hens if you want them to lay. If you can give them milk in large quantities, that will give them all the animal food necessary. Green ground bone is, of course, the best food, but it is very difficult to keep it fresh and sweet in the summer time, therefore dried bone and dried blood, or beef scrap or milk must take the place. A hen requires about half an ounce of green ground bone every day or of the dry stuff (bone and blood) half an ounce every other day. If the fowls have plenty of green food and are not laying well, give them more animal food. Perhaps your Leghorns are two years old, in which case you had better get younger fowls, as their days of greatest usefulness are over.

How Much Grain—I have been feeding three times a day, grain morning and night and a mash at noon. I feed a good handful of kaffir corn, wheat or Indian corn in the scratch pens. I have a mixed flock; I cannot well use the dry mash. How much of the grain should I give if I only feed once a day? I have fifty or sixty hens kept only for eggs and no good way of weighing grain, so please state quantity per hen and not weight.—C. A. B.

Answer—It is a good rule to feed a quart of grain for every dozen hens, the grain to be buried in the scratching pens, so they will have to dig it out. Give all the green food, clover, lawn clippings, alfalfa, lettuce, cabbage, vegetables, that they will eat, and one tablespoonful of green cut bone for each hen, three times a week. You do not mention how you make your mash. Remember that a hen needs animal food, green food and cereals; that is the balanced ration that will give plenty of eggs at all times.

———

Broken Glass for Chickens—Have started in poultry in a small way. Have had very good success so far. However, it is somewhat of a trial to get enough gravel or grit for a good sized flock on a small lot. Now, what I want to know is, is pounded glass fit to feed hens? Two of my neighbors have advised its use in the poultry yards, but I am afraid it would act on the chickens the same as it did on foxes we used to poison with it up in the wilds of Wisconsin.—J. G. F.

Answer—Broken glass or broken crockery make a very fair substitute for grit and gravel. It should be broken not smaller than a grain of wheat and have three sharp edges or corners to each piece. In using glass, be sure not to take pointed pieces like slivers, because they may pierce the crop or gizard. For several years, when I could not get grit, I used broken crockery for the chickens, and I know it does well.

———

Substitute for Green Food—Will you kindly tell me what would be the quickest and best vegetable for green food I could grow for my poultry? I planted a patch of white clover, but it does not seem to grow at all. Is alfalfa meal a good substitute where green cannot be had?—G. K.

Answer—An alfalfa patch is a good thing to have for poultry, but if you cannot have either clover or alfalfa, plant for the little chickens, lettuce, and for the older ones, kale, swisschard, cabbage, beets, etc. These in the order in which I have mentioned them are the best foods that I know of. You, of course, must judge what will grow best in your section. Alfalfa meal is a very fair substitute for green food, but of course does not come up to the crisp, succulent, fresh-growing greens.

LACK GREEN FOOD—I have three pens of White Plymouth Rocks and what bothers me is I only get from four to six eggs from them. They all look fine. I think they are rather fat. As to feed, I give them a small handful of grain in the morning in deep straw, either wheat or barley; about eleven a dry mash—eight quarts bran, four quarts middlings and nearly a quart of beef scraps; at night I give them the dry grain again. Once in a while a tablespoonful of pepper in their mash. They are not troubled with lice or mites, and have grit, oyster shell and coal before them all the time; also good clean water. Can you advise me how to feed them so as to get them down to business?—J. B.

Answer—What your hens lack is green food. At least one-third of a hen's food should be green—clover, alfalfa or some succulent vegetables. They cannot do well upon the absolutely dry food you are giving them. Add the green to your present ration and you should get eggs.

MILLET SEED—Can you tell me what makes my chickens that are from ten weeks to three months old, droopy? Is millet seed good for little chicks for the first two or three weeks? I mean millet seed alone.—Mrs. P. E. N.

Answer—When chickens are droopy it is a sign that they may have either lice, worms or indigestion. If you are feeding millet seed, that may account for it. Millet seed is very hard, round and slippery, and passes through the gizzard and intestines without being digested, and I have known of several chickens dying from it. A little used in their food may not hurt them, but an exclusive diet of millet is certain to cause trouble.

SKIM MILK—Will you kindly inform me whether skim milk is a good food for young pullets or laying hens? Which is best, sweet, clabber or curd? Is there danger of feeding too much curd or skim milk? Is curd of more value to young stock or to laying hens? I have a bunch of ten-weeks-old pullets that I am feeding clabber and bran mixed until it makes a crumbly mash. Is it a fattening or muscle or bone making ration? How would it do to feed to laying stock? I give skim milk to my laying hens in troughs which set in the sun. Will that kill disease germs or not?—L. E. E.

Answer—Skim milk is one of the best foods for chickens or hens at any stage of their lives. It can be fed either sweet, clabber or curd. By curd I mean cooked. If you cook it, be careful not to heat it above 100 degrees, or it will become tough and indigestible. There is no danger of feeding too much skim milk or clabber to fowls. The crumbly mash is good feed, but you would succeed just as well by giving them the bran dry and letting them drink or eat the milk as they want it. It is a good bone, muscle and egg making ration. I give my fowls all the milk I can spare, pouring it into troughs and leaving it till they eat it. The sun does not seem to affect it badly when it is pure milk, but if bran were mixed with it, the sun might make it ferment and then it would disagree with them.

SORGHUM SEED—Will you tell me the value of sorghum seed for poultry? Is it fat producing or an egg food, and how would it do for turkeys?—C. B. C.

Answer—Sorghum seed, broom corn seed and Egyptian corn have almost the same nutritive value. They can be fed to both chickens and turkeys with the same satisfactory results. One year when on the farm I had several tons of broom corn seed which was left where the threshers worked and the fowls had free access to it and the green-growing wheat; they got through the moult early and layed all winter, eggs galore. I never saw better laying, and the turkeys did well on it. Professor Jaffa in his most valuable bulletin (Fermer's bulletin 164) on poultry feeding, gives us the nutritive value of broom corn and of sorghum seed as both the same—1:8.4; of Egyption corn, 1:8.6; Sorghum seed is more fattening than wheat and less fattening than corn. If your fowls are on free range and have plenty of green food and animal food or milk, sorghum seed will be an excellent food for them. You should write to the Director Agricultural Experiment station, University of California, Berkeley, and ask him to send you "Bulletin 164 on Poultry Feeding," then you can see just the right way to balance your ration.

KAFFIR CORN—Is kaffir corn the same as Egyptian corn, and is it an egg food or simply a fattening food?

Answer—Kaffir and Egyptian corn belong to the same family and are very much alike. They are both fattening grains, and I prefer mixing them with other grains, such as wheat, barley, oats or buckwheat.

THE EGG QUESTION

EGG-BOUND—I have the White Minorcas. Have 15 hens and get from 12 to 14 eggs per day. I have a pullet and an old hen that seem to droop and sit around all day, and sometimes stagger; they had been laying all the time and their combs are still red, but they do not lay now. I feed them bran mash in the morning with alfalfa meal and egg-maker, and once a week chopped onions and red pepper, and at noon we give them green grass, and at night wheat, besides this they get lots of meat scraps from the table; they have oyster shell and grit before them all the time. They have not eaten anything since they felt this way, but seem to kind of gasp for breath, and they do not seem to have anything in their craws. Thanking you in advance for a reply, I remain.—Mrs. J. W. S.

Answer—Your hens certainly have been doing very well. Minorcas very often get egg-bound, as their eggs are so large they have difficulty in laying them. This may be the case with yours, and I would advise you to examine them. You might also give them some Epsom salts, half a teaspoonful in a tablespoonful of water. If it is indigestion, the Epsom salts will help that. I think your hens may not be getting green food enough.

Egg-bound is most common in sluggish birds, or those closely confined without opportunity to exercise. Active fowls, such as Leghorns, seldom take life easy enough to get fat, hence are not subject to this disease, which is largely owing to an overfat condition of the entire system, in which the egg passage is pressed upon by the accumulation of fat, hindering the passage of the egg. Not only are there large collections of fat in the abdominal cavity, but much of the muscular tissue is replaced by streaks of fat. This weakens the muscles of the egg passage, so that the egg may be arrested in the passage where it sets up inflammation. This same egg-bound condition sometimes causes death from heart disease. The bird goes on the nest to lay, strains violently to pass the egg, the heart muscles are decidedly weak from fatty degeneration, the extra exertion is too much for the weakened heart, and it gives out, the bird being found on the nest dead.

In the early stages, when the irritation is slight, it is sufficient to inject a small quantity of olive oil and gently manipulate the parts. Afterwards give cooling green food, and if the hens are too fat, reduce the ration. In case the expulsion of the egg cannot be obtained by the injection of oil, immerse the lower part of the body in water, as warm as can be used without injury, and hold it there half an hour or more, until the parts are relaxed. Then inject oil and endeavor to assist the bird by careful pressure and manipulation or by gentle dilatation of the passage.

IT CURED THEM—How long can eggs be kept for setting and do they require any special treatment? I have a favorite hen and I want to set as many of her eggs as possible, but I do not know how long they will remain fertile, as I have no hen wanting to sit at present. Several of my fowls had a touch of roup and I tried a remedy that you gave (castor oil, camphorated oil, kerosene, turpentine and a few drops of carbolic acid) squirted up her nostrils. I also mixed another remedy that you gave (cayenne pepper, mustard, vinegar, lard and flour) and gave it to the fowls, in pills, as you said. I happened to leave it where they could get at it, and found that I need not give it in pills, for they were eating it with relish. I have made the mixture several times since and they seem to be very fond of it. Their combs have become very red and although they are moulting, they are laying well. Would you advise allowing them to eat all they want of it? They are entirely well of the roup.—Mrs. H. A. H.

Answer—In reply to your first question, it is well to remember that the fresher the eggs you set, the stronger will be the chicks. I have always set them as fresh as I can get them, and I never sold eggs over a week old for setting. However, I have kept eggs from a favorite hen for three weeks and had a very good hatch. To keep them, I always lay the eggs on their side on sawdust or on grain (oats or barley) to keep them from rolling, and I turn them every day. By this means the yolk does not adhere to one side, and I have a good hatch. Some advise standing them on the small end, but it does not succeed as well as my way. I am glad your fowls have gotten over the roup. I would not advise you to let them eat their medicine, because that remedy is a very powerful stimulant, and although excellent for a cold, often curing it in one day, it will prove an irritant if continued too long. It is even now stimulating the egg organs and digestive organs greatly, as is shown by the comb, and I advise you to discontinue it, increasing the animal food; and, as yours are Rhode Island Reds, I would advise adding some oil cake (linseed meal) to the food. This will help to give a fine gloss to the new feathers.

———

SOFT SHELLED EGGS—Having read a great deal of your advice, I will ask of you a favor. Would you please tell me what can be the reason chickens lay unshelled eggs? They sometimes drop them while on the roost or out among the brush. Mine have been very bad of late; I get as many as three or four a day, sometimes, from about thirty hens. I should be real thankful to find out what to do for them.—Mrs. L. E. L.

Answer—Soft shelled eggs are not exactly a diseased condition, but may be a symptom of approaching danger. It is usually due to a lack of shell-making material in the food, or to inflammation of the shell-forming chamber of the egg duct, which no longer secrets calcareous matter. Overstimulation of the egg or-

gans by the use of pepper or stimulating egg foods, will have this effect. Worms in the intestines may also produce the irritation that will affect the oviduct, and an overfat condition will increase the tendency to laying soft-shelled eggs. This is the common cause of soft-shelled eggs.

Treatment—Provided the cause is an overfat condition, it can be remedied by giving a ration low in fat-producing elements. Give the fowls plenty of shell-forming material, such as crushed oyster shells and grit, cut bone and green food; make them work for the grain, which should be wheat in preference to other grains. One heaping teaspoonful of Epsom salts to a pint of drinking water, kept before the hens for a day twice a week, will help remove the layers of fat. Feed a properly balanced ration and do not try to increase the egg yield by using stimulants that irritate the organs of reproduction. Take freshly-crushed oyster shell and sift through a rather fine sieve, giving the coarse part to the fowls and the fine use one teaspoonful in the mash for each fowl every other day.

———

BLOOD SPOT ON YOLK—I have 150 Brown Leghorn pullets just starting to lay, and I supply a few customers with eggs and they have been complaining of finding a little blood spot on the yolk. I have plenty of nest room, so they are not crowded. I have been picking 70 to 80 eggs a day. They have abundance of green feed. I feed soft feed in the morning, wheat at mid-day, corn at evening, so if you will please let me know what the cause of this is. I will be very much obliged, because my customers are getting dissatisfied.—W. W. M.

Answer—The small blood clot you describe results from a slight hemorrhage which has generally occurred in the upper two-thirds of the oviduct. Such hemorrhages are the result of great functional activity and congestion of the blood vessels. They are excited by any of the causes which lead to congestion and inflammation and are to be counteracted by green feed and less animal food and by the suppression of red pepper or any stimulants. Give a little Epsom salts in the water and add about twice the amount of salt you are giving to the mash in the morning, leaving off the red pepper.

LARGEST WHITE EGGS—I am starting or trying to start a poultry ranch and would like to ask you a question recently asked by some one else; but in a little different way. Which of the good laying breeds lay the largest white eggs? My aim is for good city trade.—E. A. M.

Answer—The Black Minorcas have the reputation of laying the largest white eggs. The White Leghorns are their close competitors. It very much depends upon the strain or family. For instance, one set of fowls may have been selected for beauty of feather and form and their owners may not have chosen those that layed the largest eggs, whilst some have carefully chosen the largest egg-layers, and bred from those, not caring for exhibition birds, and again a third party might have united these two qualities and have both prize winners and the best of layers. It depends upon the ability of the breeder and also upon his object.

Black Minorcas do admirably in the climate of Southern California. I do not know how they would grow in a damper, colder climate. You would have to inquire of people who have had experience in that kind of a climate.

———

SUDDEN DEATH—Lately I have had three hens die suddenly, and apparently without cause; my neighbors have also lost several. Perhaps you can enlighten us and suggest a remedy. The hens were laying, combs red and large, crops full of wheat, etc., but die on the nest over night. I held a post mortem examination and could find nothing radically wrong. Each had well-formed eggs and many of them. They roost high in the open air; run out nights and mornings on alfalfa. I feed wheat mostly, and once every other day, hot bran mash with a spoonful of egg-maker. Have had over 40 dozen eggs without interruption since January 1st, from twelve pullets—Minorcas—of my own raising. This is the first death I have ever had, except of the little chicks. Pens are clean, no lice or mites. Have studied closely and can't "savy." Perhaps you can. The heart of the first one seemed the only cause for death, as it had a large inforct, probably fatty

degeneration; the other was normal.—Dr. J. A. B.

Answer—I think, as your hens died on the nest, that they had some difficulty in laying, and were probably egg-bound. The Minorcas laying a large egg, are frequently subject to this trouble, more so, in fact, than the other breeds which lay smaller eggs. Straining in laying frequently is the cause of a blood vessel breaking in the head, which, of course, results in apoplexy. Minorcas rarely suffer from an overfat condition, as they are a very active breed.

———

EGG-EATING HENS—Would you kindly tell me how to treat egg-eating hens? What will cure them?—Mrs. R. E. G.

Answer—The best way is to cut the head off the offender and eat her, for she is certain to be fat. The information you ask for is as follows: Mr. Morse (a chicken expert) gives five remedies for the bad habit of egg-eating. First: Fit up an arrangement whereby the eggs, as soon as layed, slide down and out of sight, into a sort of false bottom under the nest. The hens will not eat them because they cannot get them. Second: Have a lot of China eggs lying about promiscuous-like on the floor. Trying to eat such eggs is likely to discourage egg-eating. Third: Fix up a hollow egg with aloes. One bite is enough. Consult the corner druggist as to how to make the mess. Fourth: Have grit and crushed oyster shells about in abundance in self-feeding boxes. Fifth: Do not stuff your hens full of mash in the morning and let them sit around all day, like "Father" in the song, "Everybody Works But Father," but feed them grain in litter and make them hustle all day. This keeps them out of mischief. Mr. Morse's advice may be good, but I recommend using trap nests by which means you will easily discover the guilty hen, and if she is not too valuable, the verdict should be decapitation. Keep your oyster shells, grit and charcoal before your hens and there will be very little egg-eating, for it is a vice which always commences with weak or soft egg shells.

HATCHING WITH INCUBATOR AND HEN

POOR HATCHES—We have been running our incubator since February and our hatches have been quite poor. Our hens are two years old and so are our roosters. The hens are fed regularly, and have a large run with plenty of alfalfa; a clean airy coop.

The chicks, when hatched, are strong and vigorous. We have some six weeks old, and we have not lost one, but when they are hatching many die in their shells. Out of 450 eggs 77 tested out not fertile or dead germs, and out of 373 remaining eggs, only 182 hatched. We are hatching White Leghorns. Can you tell us what to do, or what the matter is? We have been following your advice in many things.

Do you think that slamming of doors or jarring is bad for incubators when hatching?—Mrs. M. F. DeW.

Answer—I think the fault in your incubator is that it has not sufficient ventilation. An insufficiency of oxygen will cause poor hatches such as you describe. With the care you give your fowls and their being two years old, the fault does not lie in the parent birds or their eggs, therefore it certainly comes from a faulty incubator. In the future, air the eggs three times a day; fan out the stale air of the incubator each time you air the eggs, and if you find they are drying out too much, sprinkle them, after the first week, twice a week with warm water. Slamming the door or jarring the incubator during incubation is not advisable, but on the day of hatching it would not injure them.

INFERTILITY—Will you kindly tell me what to do to make eggs more fertile? I have a fine pen of Columbian Wyandottes, eight pullets mated with a cock two years old. They are fed on dry, mash of bran, ground barley, corn meal, alfalfa meal and beef scrap, with plenty of grit, shell, charcoal and ground bone before them all the time, and are running in a corral of grass and clover; they have plenty of fresh water and the hens lay well. What chicks I do get are strong and healthy; out of fifteen eggs only two were fertile.

I have another pen, four hens, two years old, mated with a cockerel one year old. Fed the same in every way; their shells are smooth but full of clear spots. What shall I feed to make shells better?—Mrs. E. H. G.

Answer — The usual requirements missing from the food when eggs are infertile are green food and animal food. therefore, I would advise you to feed more green food, more animal food and a great deal less barley and corn meal. Wyandottes are apt to get too fat to have good fertility unless they have plenty of exercise and the four old hens require more lime. Mix some fresh quick lime in water to the consistency of pancake batter; let it stand 24 hours. then pour out a cake of it on the ground. It will soon dry, and by crumbling a little of it every day, the hens will pick it up. Add a teaspoonful of baking soda to a quart of their drinking water and keep this before them for a week. By this means I think your egg shells will improve.

CRIPPLES — Some of my incubator chickens are almost cripples when they are taken from the incubator. Some have crippled, crooked and crumpled up toes, others have one leg too short, or turned out the wrong way, and some of them are not able to stand up—they hold their head back so far that they fall backward.—A. H. S.

Answer—The cause of cripples invariably is irregularity of temperature in the incubator. Your incubator has been too hot at some period, probably the last week; this causes cripples. Those that hold their heads back do so from the eggs not having been turned sufficiently during incubation.

As you do not mention the name of the incubator, I cannot tell you just where the lack is, it may be poor oil; it may be it is run in a draught and it may lack ventilation.

LACK OXYGEN—I took 200 thrifty chicks from the incubator about eight weeks ago. They did very well for about two weeks, when they began to die and today I have 50 left, and these look too scrubby to be worth raising. I have given them extra attention and the best feed. They get pale around the head, grow weak and are skin and bone when they die. I think they have consumption. The brooder is a tight box and no ventilatiotn, except the lid has a round hole about as large as a teacup, and the little entrance window about six inches square. An iron pipe running through is the heating arrangement. Inside the box, to fit close over the pipe, is a cap of wood with flannel curtains dropping to the floor under which the chicks hover. Don't you think this is too close a place? The outside box is only 6 inches deep, then they hover inside; this only gives 4 inches space for the chicks. Please tell me if you think the lid to brooder would be better of wire or where do you think the trouble is? Also tell me how granulated milk is prepared. We have lately begun feeding to everything in the poultry yard, beef scraps, bone meal and linseed meal in what we think proper proportions once a day. Should chicks only eight weeks old be fed this ration the same as hens? What causes eggs to be ridgy and uneven? Can one feed to produce larger eggs? Our hens are large, but lay small eggs.—Mrs. J. B. S.

Answer—I think that the lack of oxygen in your brooder is the only difficulty with your chicks. Still, I am very much afraid that tuberculosis may have got in and infected the brooder. If possible, move your chicks into a weaning house, open entirely on one side (or only closed with chicken wire). Make a little frame of gunny-sacking or out of a piece of blanket that they can go under. This will rest upon their backs to keep them warm. Give them no other heat. At this season of the year (August) eight weeks old chicks should have no heat whatever, at night. I think you are keeping your chickens too warm, without enough fresh air and possibly they may have mites or lice. Air their sleeping place well; put the hover out into the sunshine every day. This will kill the germs of tuberculosis better than anything.

Granulated milk is made at Binghampton, N. Y. I do not know the process.

Chicks eight weeks old can have the beef scraps, bone meal and linseed meal in the same proportions as hens.

Uneven eggs are caused either from defect in the oviduct or from an insufficiency of lime or hurried laying.

Some strains of hens lay small eggs and overfat hens will lay small eggs. More protein added to their food will often increase the size of the eggs: By choosing the large eggs for hatching, you can increase the size of the eggs in the next generation.

CHICKS DYING IN SHELL—A large per cent of my chicks, fully developed, die the day they are due to hatch, even after pipping the shell. They seem to dry in the shell.—Mrs. D. D.

Answer—Float the eggs in warm water. That will help the chicks to break through the shell better than anything I know of. Next time try sprinkling the eggs after the eighth day twice a week with warm water. I think you will find it is what is needed in your dry climate, and is likely to help matters.

FOOLING THE HEN—Is it possible to fool a sitting hen into caring for some incubator chickens when she has not hatched them herself—Mrs. C. R.

Answer—If your hen has been sitting for a week or ten days, she will "take to" the chicks as well as though she had hatched them herself; especially if she is a Plymouth Rock or Buff Orpington. Those two breeds have a greater affection for chickens than some of the others. Be sure that the hen is entirely clear of lice, and if she is a large hen, put from 15 to 18 under her at night; a smaller hen should have from 12 to 15, not more, if you expect the chickens to do well. I have trained capons to act as mothers; they do even better than the hens.

THERMOMETER—Will you kindly tell me where I could get tested thermometer for incubator; also where I could have one tested which I already have? —H. H. C.

Answer—At any good drug store you can have your thermometer tested. If you want to buy a new one, go to the agent selling your make of incubator. Take the new one also to the druggest and have him test it thoroughly, because the thermometers, as they are seasoned sometimes vary some degrees, and even a new one cannot be trusted.

HELPING THEM HATCH—I find my White Plymouth Rock eggs are very slow about hatching and some I know would die in the shell if I had not dropped a few drops of lukewarm water on their heads, as it seemed they would get about half out and then the white skin would dry on their heads and hold them fast. After having two die in the shell, I found they would free themselves if a few drops of warm water were sprinkled on them. I kept moisture in the pans all three days and part of the fourth and they are still slowly hatching. This is the twenty-third day. Do you think I should keep the moisture pan full for a week—I mean the last week of incubation? Please send me an idea on chick feed, as I cannot get good, clean chick feed here.—Mrs. P. W. B.

Answer—If you had only mentioned the name of the incubator you are using, I could have better diagnosed your case. As it is, all I can say to you is to follow the rules and directions they give you as closely as possible. With some machines it is very advisable to sprinkle the eggs twice a week after the first week with warm water; this seems to make the shells more brittle and prevents the inner lining skin from toughening. I have found this better than keeping much moisture in the machine. The moisture in the machine seems to make the chick grow, but does not make the shell brittle. Your Plymouth Rock eggs should hatch promptly on the 21st day. The delayed incubation indicates that part of the time the temperature has been too low. Are you sure that your thermometer is perfectly correct; have you had it tested? On the efficiency of the thermometer much depends. Many thermometers that are accurate at first become, through the use of unseasoned glass in their manufacture, absolutely incorrect after a few months' use. Others are really only within two to four degrees of being correct, therefore, be sure you have your thermometer tested. About the chicken feed, write to the Experiment Station, University of California, Berkeley. This gives you the list of foods available in your part of the country, with the proper proportion for mixing them, see page 38.

EGGS FOR HATCHING—Will you kindly tell me what is the matter with my eggs? They will not hatch well. Our hens are Brown Leghorns and Rhode Island Reds. I only got fifteen chickens in my last batch. When we break the eggs after we know they will not hatch we find the chicks dead, but fully formed and just ready to hatch. Perhaps the shells are too hard. Will you please tell me what to do to make a softer shell? Feed according to your directions.

Is it necessary to put moisture in the incubator? Does it hurt the eggs to sprinkle them with warm water if we think the shells are too hard? I will be very thankful if you will answer this, as I want to know before I commence to save eggs for next incubator lot. I do not keep them over two weeks and keep them in a cool, dark place, turning them every day.—Mrs. G. A. M.

Answer—I wish I could tell you for certain what causes chickens to die in the shell. I have my theories about it, and I believe it comes from the eggs not being aired and cooled sufficiently. Cooling them and then warming them up again seems to make the shells more brittle, and this is the same under hens. If I notice that a hen is sitting too closely, I take her off twice a day to cool the eggs. With an incubator I would air them and turn them three times a day, and either sprinkle them three times during the last ten days or float them in warm water two days before the hatch is due. Float them from three to five minutes, and then put them back into the tray while they are wet. I do not believe in putting moisture into the incubator unless the directions call for it.

INCUBATOR CHICKS DYING OFF—We have started in with the R. I. Reds, and have been fairly successful until our last hatch. Out of 65 eggs 44 came out. Last Saturday they commenced dying off, just fell seemingly from weakness and died soon after. We have fed them chick feed, bran, Indian meal, cayenne pepper, beef scraps, twice per day, and a little germazone in water occasionally. —C. R. H.

Answer—From your description I am afraid that the chickens have either been chilled or may have been overheated. Either one of these conditions will cause the symptoms you describe. All you can do now is to give them rice boiled in milk, adding a tablespoonful of ground cinnamon to each pint. Give them also chopped lettuce and onions. Do not give any corn meal or beef scraps. When chicks have been overheated either in incubator or brooder, it so weakens their bowels that they cannot digest their food and they die of starvation.

POOR HATCHING—I should like very much if you can give me some information about my hatching eggs in an incubator. I bought a new incubator this spring. I have set it twice and had the same results both times. The chicks form fully and then most of them die in the shell. As the same eggs do fine when put under a hen, I think it must be that I make some mistake in my treatment of the incubator. I have, as nearly as possible, followed the instructions that came with it. If you can give me any assistance, it will be appreciated very much.—Mrs. W. D. W.

Answer—Your incubator is a good one. Its fault, for they all have some little fault, is that the ventilation is insufficient. Take the eggs out and air them after the first week three times a day. This will counteract the lack of ventilation. This cooling and then heating up again of the eggs makes the shell more brittle, so that the chick is able to break its way out much more easily. Another thing I found in using that incubator is that by taking the middle eggs out of the row, one in each hand, and putting them at the end of the row, and then pushing the others along into the vacant places, I got a ten per cent better hatch. I got the idea from Egypt. Of course, you must be sure the machine stands level and that the thermometer is correct.

———

TROUBLE WITH INCUBATOR—I want to ask your advice about our incubator. We bought it new in January. Out of 200 fertile eggs we got 75 chickens, and all but nine died before they were 10 days old. We thought it was the fault of the brooder. There were many cripples among them, but they all died of bowel trouble. On April 30th we hatched 117 out of 150 fertile eggs, and gave the chicks to old hens, as we had laid our previous trouble to the brooder. But now the last are going the same way. Chicks hatched under hens at the same time are healthy and strong. We have only lost one so far. We feed prepared chick feed and take the best of care of the chicks. The incubator runs perfectly, always 103, until the chicks begin to work out of the shell, when it runs up to 104 and 105. We have set the incubator again. It will hatch May 29th. We do not intend to give up.—W. S. R.

Answer—The trouble is in the incubation. At some time or other the heat has been too great. This is shown by there being cripples. I know it, because

I have had the same experience several times myself. Once a hat was thrown on the machine; just touched the regulator; was only on for half a day. Another time a newspaper did the same thing. My big cat slept on the incubator another night and lost me the hatch. Each of the times I worked with the little chicks, giving them everything I could think of, but without saving them. Now, I think there is a possibility that your incubator does not stand level and that, therefore, one side or corner of the machine is a very little higher than the other. That side or corner would be hotter than the other side without affecting the thermometer and would cause all or most of the trouble. Again, are you sure the thermometer is correct? Borrow the doctor's clinical thermometer. This is what I did, and put them both into a bucket containing about two quarts of water at 103 degrees and compared the two. You do not mention if the hatch came out on time. I feel sure that the eggs have been overheated, or part of them have, and in this way the bowels of the chickens have been weakened, the yolk of the egg has not been digested, and they have dwindled and died, or bowel trouble has come on from the undigested yolk putrifying inside of them. I have made so many post mortem examinations that I feel sure of what I am telling you. Examine your incubator with a spirit level to see that it is level. Test your thermometer and then try again, at the same time setting one or two hens, and as incubation proceeds examine the eggs, comparing them. I think you will find that the eggs under the hen dry out quicker than those in the incubator. However, if this is not the case, if your incubator eggs dry out too quickly (the air space being larger than that under the hens), you will have to regulate this by the ventilators of the incubator. Keep them closed. As yours is a hot-air incubator, there is no need of fanning out the stale air. The fault, if any, with your incubator is too rapid a circulation of air, thereby drying the eggs out too soon. I think you had better run it half a degree cooler than you have been doing. I say this because the cripples and bowel troubles denote too high a temperature. I hope these hints may help you. Let me hear from you again if you have any more trouble.

———

NATURAL INCUBATION—I am a reader of your articles and get much good from them. Am a beginner and have a great deal to learn. Will you kindly answer the following questions:

1. Should a setting hen be shut on the nest and be let off each day? If so, how long should she be allowed to stay off the nest?

2. Do the eggs get enough moisture in natural incubation?

3. Is it good to sprinkle the eggs with water? If so, how often and in what stages of incubation should this be done?

4. How long should chick feed be fed to chicks, and what is best after discontinuing this food?—R. M.

Answer—It is best not to shut a hen on the nest, but to allow her to get on and off as she pleases, unless there are other hens that can get to the nest to disturb her. It is a good plan to take the hen off the nest at a regular hour every day. I prefer about five o'clock in the evening, as then she will go back before supper time. A hen can be off the nest in pleasant weather from twenty minutes to half an hour. She should be allowed to stay off long enough to eat all she wants and to dust herself. It is necessary for her to come off at least once every twenty-four hours.

2. Eggs usually get moisture enough from the perspiration of the hen. I like to float the eggs in warm water two days before the hatch comes off. I think it helps the eggs to hatch well and it also shows, by the eggs bobbing about on the water, which eggs have live chicks in them.

4. Chick feed should be fed about six weeks, but it is best to begin when the chicks are three or four weeks old to add wheat and kaffir corn to the chick food and make the change gradual. Commence by one-fourth of the larger grains and three-fourths of chick feed. Then gradually increase the kaffir corn and wheat until that is the principal feed.

BROODER CHICKS—I shall have to come to you for help about my little chickens, as I know that you know what to do.

I am only a beginner. I have an incubator and hot water brooder, and before I bought your book I could not make them hatch, but now, with its help, following your directions, I have a fine hatch. I turned and aired the eggs as you said. Now my chicks (White Leghorns) are two weeks old and I have lowered the temperature in the brooder

about one degree a day; but about every other day one will die. I have thirty-two in the brooder, so they are not crowded at all. I have put insect powder on them and they are fed chick food; they have plenty of fresh water in a fountain, which I keep in their yard. I make them work in alfalfa for their feed, as you instructed. They are not stuck up behind, as far as I can tell, but when one is about to die, it goes up into a corner of the brooder under the pipe.

If you will give me advice about what to do, I shall be very much obliged, as I am afraid I shall lose them all.—N. H. H.

Answer—I am glad you had a good hatch. The fault with that incubator is lack of ventilation, and of the brooder is that there is a draught on the floor, so that the chicks' feet are cold. I tried a good many plans with that brooder, and finally I built them over. However, the best plan before I changed them I found was to put on the floor a gunny sack or bit of warm old carpet, and on that put nearly two inches of chaff or finely cut straw or hay. I also left the lid a little bit open. Before that the chicks' heads got too hot on the pipes and their little feet too cold.

I am rather surprised that they have not been troubled with diarrhoea.

FAULTY INCUBATION—I am a beginner in the poultry business and would like to ask you a few questions that have been troubling me:

1. I have been hatching chickens and ducks in an incubator and they don't hatch as well as with a hen. I find quite a number dead in the shells. I do not understand it as I follow the directions that come with the machine.

2. A number of the chicks "walk around on their knees." Some of their legs stick straight up and they flop along on the joint with the aid of their wings. They soon die. Why is this? Is there any way to avoid it?

3. I had twenty ducks hatch with hens and have only eleven left. We first notice them to lag behind the rest, then as they grow more stupid they fall over with their heads thrown back as people do when they have spinal meningitis. Can you tell by this description what was the matter with them?—L. B., Corcoran.

Answer—The trouble is that the heat has been irregular in your incubator, and probably the eggs have not been aired sufficiently.

2. Cripples, such as you describe, invariably come from overheating, especially the last ten days in the incubator. It may be only for a few hours. It is such a pity, for it always seems to be the biggest and best chicks. I have once or twice succeeded in straightening out the legs and setting the knee, fastening it with a rubber.

3. The trouble with the ducks is severe indigestion. It may be they have not had sand enough in their food, or they have eaten some animal food that was not fresh—was decaying. Lack of shade will give the same symptoms. The drinking vessel must be deep enough for them to get their entire bill under water, for they require to rinse their nostrils many times a day and will die if they cannot.

BROODERS—(Mrs. S. M. G.)—I would like to tell you about the brooders I made from your description of them. I have used the Fireless Brooder for five months and have had no trouble in getting the chicks to go inside when they are cold. When I first put fifty chicks into the Fireless, the weather was cold and at first I found, like others, that the little fellows did not know where to go when they felt cold, so on the third day I put a gallon jug of hot water in the center of the brooder, covering the jug with a hood made of several layers of newspaper. I took two or three chicks and held them against the jug until their happy chirping brought all the others; after that I had no trouble. I removed the jug at night and put it back in the morning for a few days, filling it with less warm water each morning. During the summer months I did not find it necessary to put any attraction in the brooders as the chicks seemed warm enough from the first to spend the entire day in the sun.

This account from Mrs. G. will interest and help many of our readers.

YARD ROOM

How Many Chickens to Keep on a City Lot—Will you kindly tell me how many chickens can be kept on a city lot seventy-five by a hundred and eighty feet? Do you think chickens will lay well during the rainy season in Seattle, Wash., if they are properly fed and housed? How big a house do we need for fifty chickens?

Last September we bought thirty Plymouth Rock hens and thirty pullets. We got from ten to sixteen eggs from the hens per day, until about the middle of December, when they began to fall off. We are still getting that amount, but half of them are from the pullets. Do you think they are doing as well as we could expect?—Mrs. L. E. S.

Answer—In your climate it would very much depend upon the shelter from the rain that you can give the chickens. Fifty chickens should be divided into two pens with two houses. Each house not less than ten by twelve feet in size. I would advise a good scratching pen to be made either adjoining the house and covered with a roof, or else make the scratching pen to extend underneath the dropping boards. You might keep several hundred hens upon land 75 x 180 feet, if you have ample house room for them, so they would be well sheltered from the rain. Hens that are wet every day will not lay well. Your fowls are doing well, considering the wet weather you are having.

How Many on Two Acres—I have two acres of land, of which I will have a hundred feet by one hundred feet for an alfalfa patch, the rest for chickens to run around and have the patch for them to feed on for an hour or so before going to roost. Kindly let me know how many chickens I can raise on the two acres at the most.—M. J. P.

Answer—I think you can keep a thousand chickens on your two acres. You must be careful not to have more than fifty to roost in one house. It is the crowded condition of houses at night that brings trouble and disease. Be sure to give them shade during the day and plenty of good fresh water, besides, of course, the balanced ration. Allow them two hours a day on the alfalfa patch.

Five Acres—Will you kindly tell me how many White Leghorns I can successfully raise on five acres of land? I want to grow alfalfa and some vegetables for feed.

Will you also tell me if I can hatch turkeys in an incubator?—J. W. L.

Answer—You can raise a large number of Leghorns on five acres of land. I know one party that has 3000 Leghorns on three acres, but it entirely depends upon knowing how to do and doing it right. Better begin with a small number and when you succeed with those increase your flock.

Turkeys can be hatched in an incubator and raised in a brooder, but must be kept entirely separate from chickens or they will die.

YARD ROOM—I want to raise about 60 pullets for next winter. I have about a hundred chicks hatched out. All the yard room I can spare is on a town lot about 50 x 75 feet. Do you think this would be enough room for them?—Mrs. J. F. Y.

Answer—It all depends upon the care you give them; if you can supply them with shade, plenty of green food, clean water and a good scratching place and the proper food, it will be plenty large enough. Be sure to keep them clean and free from mites and lice.

BURGLAR ALARM—I refer to the mention made by you of an electric burglar alarm to protect poultry houses, and would venture to inquire whether such an alarm may be installed by one not a professional electrician. Upon what principle is it based, and what are the materials needed?—H. M.

Answer—I put in the burglar alarm you speak of myself. I am not a professional electrician, but I went to the electrical supply house, bought from them the ordinary alarm fixtures which are used at the door and windows of residences; they explained to me how to set them, and I did it by their directions. I did not find it difficult. None of the doors or windows in my hennery could be opened four inches without the alarm gong at the head of my bed ringing. I should think you would have to understand a little about it to put them in.

MATING AND BREEDING

AGE FOR MATING—I wish to ask if a cockerel should be mated after he attains a year in age or can he just as well stay till a year and a half or two years old before being mated?

Also I wish to know if it is quite as advantageous to mate a rooster with a pullet of his own clutch, supposing the pullet and rooster are both a year and a half old. I would like to do that if you think it advisable.—M. S. H.

Answer—The earliest age at which a cockerel may be mated should be about ten months, not earlier if you want large, vigorous chickens. I consider the best age for getting sturdy chicks is for both parents to be about two years of age. You can keep a male bird as long as you wish without mating him, but he should be entirely out of sight and out hearing of the hens, otherwise he will fret to get to them. I have known several to drop down dead from getting too much excited at seeing other young males in the pens with the hens.

From a year and a half to three years of age is undoubtedly the best age at which to mate the fowls, but you can have very good results with older fowls.

In your place I would certainly mate the year and a half male with the year and a half hen and expect good results, for they should both be in their prime.

MATING BROTHER AND SISTER—Is there any objections to mating a rooster with hens of his own clutch if they are all old enough, say a year and a half or two years old?—Mrs. G. S. H.

Answer—It is considered best not to mate brother and sister together, yet this is always done in making any new breed, and as yours comes from a three hundred egg a year hen, I would advise you to do so.

BREEDING—I have a nice R. I. R. cockerel. He is good shape and color, but he is not up to standard weight. If I breed from him will he produce chicks larger than himself if they are well taken care of? Is there any chance of getting perfect specimens from fowls under weight? I bought some very fine looking hens, but their breasts are uneven. I also got eggs from the same stock and the pullets have crooked breasts. Kindly tell me if that trouble will be handed down if I breed from them.—Mrs. C. R.

Answer—As a rule, the chicks take their size from the mother. If your R. I. R. hens have a good size, the chickens will be larger than the cockerel, if you feed them for large frame. If the hens are under weight and size, you may have difficulty in increasing the size of the offspring. Some people think that crooked breastbones come from chickens roosting on a narrow perch when they are young; however, I think it is generally conceded that crooked breastbones are often hereditary. You will know if your chickens have roosted at too early an age. If not, it is hereditary and you had better change the strain.

MISCELLANEOUS QUESTIONS AND ANSWERS

SHIPPING YOUNG CHICKS—Do you think I can order eggs incubated 31 miles from here and have the young chicks sent by stage with perfect safety?

We are feeding corn of our own growing which is quite musty. I have been afraid of it, but so far cannot see that it has hurt them, although yesterday a hen sat around all day droopy like. I wondered if the musty corn affected her.

Last summer I brought into the house some small chicks that seemed about to die, and seeing they had lice, I dusted them thoroughly with buhach. The lice soon dropped off of them, but the chickens died. Can too much powder be put on them?—Mrs. C. S.

Answer — Chickens could travel a thousand miles before they are twenty-four hours old if packed in a box carefully. That is, of course, before they are fed. Last year I sent some from Los Angeles to Berkeley. They were out 36 hours, but arrived in perfect condition, all vigorous and ready for their first meal in their new home nearly a thousand miles away.

Musty wheat or corn is very unwholesome for chickens. Buhach would not kill the most delicate chicken or turkey, but is death to all insect life. The chickens were doubtless dying before you powdered them.

CASTOR BEAN BUSHES—I have been thinking of planting castor bean bushes in the chicken yard for shade, but was advised by a neighbor not to do it, as the beans would drop off and if chickens ate them they would be poisoned. Would like your advice, please. The bushes grow quickly and make good shade, so would like to try them. Do you think it would be O. K?—J. H. S.

Answer—Castor beans are poisonous to both ducks and chickens if they eat them, so I would advise you to plant something else. Get cuttings of fig trees, about ten inches long, bury the whole length except one inch, water well, and you will have shade in a few months and fruit in two years. I find figs excellent in the chicken yard, and the chickens do not eat the leaves and bark. Would advise planting also other fruit trees and the quantities of fruit you will have will soon repay the trouble. In the meantime you might plant sunflowers. They make good shade and their seed is excellent food for the chickens.

CAPONS—Will you kindly give us an article on capons? What is the demand for them, if any? What do you think of the difference in profits between them and broilers? If there is any truth in the statements published in regard to capons in the Eastern markets, they ought to be money-makers here. Am fitted for the business, but desire more information in that line before attempting much. I think the R. I. Reds would make extra good ones, and I should like marketing mature birds instead of those a few months old. Capons for the Philadelphia market have to be a year old to command the best prices.—H. J. K.

Answer—Capons bring a good price now in Los Angeles, especially if you can make a contract with some of the large hotels for them. This you can only do by having a large and regular supply. The price last year was from 30c to 35c per pound, which is a paying price. Broilers pay about as well when you take into consideration that you can turn them off at eight weeks of age. This would be your better plan, as you are limited for space and you would not have the expense and trouble of carrying them for another ten months. I would advise you to sell as broilers all the young males you do not wish to keep for breeders. This will give you more room for the pullets and you need space to have your pullets develop well for the fall and winter egg market. Capons are, undoubtedly, money-makers for those who have plenty of space, and where food is cheaper than it is here this year. Personally, I found that capons did not pay as well as roasters. These were young roosters that were about eight months old and that I milk fed. I found I had to keep my young males until I could see how they would develop. I began by caponizing, but being economically inclined, I found the milk-fed, uncaponized, eight-months youngsters paid me best. Since then the market for capons has improved here, and if you had more room and could buy up young cockerels, caponize them at about three months of age and turn them off in the following spring, just when turkeys go out, you might make some profit on them. It has been found that the Brahmas or crosses of the Brahmas are the best for capons .

FROM FAR AWAY ALASKA—Commencing with the first of March for the last three years my chickens begin to lose their feathers in front of their neck. I feed them wheat, corn, shorts, cooked potatoes and cabbage. They have no lice. I also give them plenty of charcoal and grit. I have a chicken house 30 x 30, logs with moss between, lined inside with shakes. I also keep fire in a stove to keep out dampness.—H. C. C., Sumdum, Alaska.

Answer—Not knowing your climate, scarcely like to venture an opinion about the reason for your hens losing their feathers. Your rations seem good, all except there is no animal food in it. I think you should give them fish with their cooked potatoes. Do not feel alarmed about them losing their feathers, as it may be on account of the climate.

TECHNICAL NAMES—Will you please tell me how old "friers," "broilers" and "prings" are? When is it safe to feed wheat and mash to chicks?—Mrs. M. N.

Answer—It is not by the age that we decide upon the size of the chickens, or their names. "Squab broilers" weigh one pound and are usually from a small breed, fattened as quickly as possible, the age being about six or seven weeks. "Broilers" weigh from one to two pounds, the age being about eight weeks. "Friers" weigh from one pound to two and a half pounds; age, about ten weeks. Young "roasters" from two and a half to three or four pounds, age about three months.

Feed the wheat to chicks as soon as they will eat it, commencing to add it to the chick feed. I commence also to add kaffir corn at the same time. Some chickens will eat it earlier than others; mine, a large breed, usually will take it at three weeks.

HENPECKED HUSBANDS—I cannot keep my hens from picking the combs of the roosters. Could you tell me the reason for it? Also a remedy for it? I have tried everything I know for it. I feed meat twice a week.—R. M.

Answer—This habit or vice usually comes from a lack of green food or meat in the ration. Very often the habit is acquired by imitation and thus it may be introduced into a flock by a new bird which had contracted it elsewhere, or it is spread through the flock from a bird which is led to it by indigestion or other disease of the stomach. It is sometimes started by lice. The hen sees one crawling on her mate's comb and tries to peck at it, wounds the comb, tastes the warm sweet blood and keeps up the habit. The others imitate her until the poor henpecked husband is in a sorry plight. The preventive is plenty of green food, plenty of exercise and animal food. The cure, the hatchet for the worst hens, or if they are too valuable, let them run without the male bird, only admitting him to the pen for an hour a day in the afternoon. Give the hens a good run in a grass-covered yard. Feed plenty of green vegetables; onions chopped are particularly efficacious. If the yard is small, prepare a scratching shed, covering the floor deeply with straw and scatter grain in the straw for the morning meal, so the fowls will be compelled to scratch and work to find it. Add bicarbonate of soda to the drinking water in the proportion of about 20 grains to the quart; put a small quantity of salt in the food, or nail up a piece of salt pork for the hens to peck.

PAINTING BROODERS—Will you kindly tell me if painting the brooder on the inside with crude oil will injure little chicks?

We have ordered 100 Brown Leghorns for March 15, and have got a second-hand brooder. Of course, we want it perfectly clean, as we are beginners and are striving for success. A friend of ours gave us five gallons of crude oil and insisted on our using it, but I thought it wise to ask some one more experienced. Thanking you in advance, yours truly.—Mrs. G. S. McW.

Answer—I would not advise you to paint the inside of your brooder with anything as strong as crude oil. It will do very well to paint the outside of the , hen house and the outside of the brooder house, and will last for several years, preserve the wood and keep away vermin, but is too strong for the little chicks.

I will tell you what I would do were I in your place. I would take good hot suds and a brush, either a whisk broom or a scrubbing brush, and thoroughly scrub out the brooder. If I thought there were any mites or lice in it, I would add a cupful of coal oil (kerosene) to the suds. I would then put it in the sun to dry, and when it was dry I would wash it all over—hover, felt and everything—with a solution of bichloride of mercury. You can get tablets of it very cheap at any drug store. Put about four or six tablets in a pint of water and when it is dissolved wash all over the brooders with it. Or get corrosive sublimate; have the druggist dissolve it in alcohol, and paint that over the inside of the brooder. This will destroy all germs of any disease or any vermin. This way of soapsuds, followed by the mercury, is the most perfect disinfectant you can find. It will kill tuberculosis, chicken-pox, cholera, etc., germs, and has no bad smell to injure chicks.

———

How Long?—Would you kindly answer how long after the eggs have started in the hen does it take before the hen lays? Thanking you in anticipation.—W. B. M.

Answer—As soon as a pullet is three months old there will be found inside her a bunch of tiny embryo eggs. These are called the ovaries or egg organs. If the hen is active, in good health and properly fed, these will, one after another, turn into eggs, but the hen must be fed the elements of the egg in order for her to make the eggs, and it all depends upon the food how long it will take the hen to accumulate the proper proportion of each element to make the eggs, that is, the elements of the egg rightly balanced, enough fat and protein to make the yolk, enough albumen and water for the white, enough lime for the shell, each. in its right proportion.

———

Soft Shell Eggs—Please tell me why my chickens and turkeys lay soft shell eggs.—R. A. D.

Soft shell eggs come either from an insufficient supply of lime in the rations or overstimulation of the egg organs by the use of spice or so-called egg foods. Worms may increase in the intestines to such an extent as to stimulate the egg passage to push along the egg beyond its usual distance. An over-fat hen has a tendency toward laying thin-shelled eggs.

Dr. Woods gives this advice: "Fowls kept closely confined in cold weather and not given a sufficient variety of food are apt to lay soft-shelled eggs. The trouble may be due to some disturbance of the egg organs or to improper food, careless feeding and lack of exercise. It usually responds very promptly to treatment. See that the birds are supplied with plenty of good grit and oyster shell. Feed green food, scalded short-cut alfalfa or clover. Also give cabbage, beets and turnips fed raw whenever they can be obtained. Feed a variety of good, sound grain and some animal food. The grain should be fed in the scratching pen."

———

Saw Off Long Spurs—I wish a little information in regard to a rose-comb Rhode Island Red rooster two and a half years old. He has very long spurs, which makes it difficult for him in scratching when I feed them in the scratching pen. Is there any way of taking them off?

Answer—It is very advisable always to cut the long spurs off the male birds. as they are very apt to injure the hens with them. I find the best way is to saw them off with a fine meat saw about an inch from the leg. I do not saw them close enough to draw blood. You can also file them off, but sawing is quicker, and if the edges are rough, use a small file to make them smooth.

CHICKEN MANURE—Please answer immediately. How can chicken manure be preserved, and where can it be disposed of, and at what price? Answer and oblige.—Mrs. M. A. S.

Answer—The easiest way of preserving chicken droppings is by placing dry earth or sand or kainit under the perches, sweeping this up two or three times a week and placing it in barrels or boxes. Anyone with a citrus orchard is glad to get it for fertilizing the trees. I know one man who pays $7.50 per ton for it. I do not know what the market value is, but I know that it is considered worth just four times as much as stable manure and that it is a most excellent fertilizer.

———

FIRELESS BROODER—I make bold to ask you for a little information. Will you kindly tell me of the fireless brooder? Can you give me the plans for constructing one, or tell me where I can get the plans? Can little chicks just hatched be put in the fireless brooder?—Mrs. W. W. G. Arizona.

Answer—Take a box about ten inches deep, and from a foot and a half to two feet square. Rip the box six inches from the bottom to four inches from the top, so there will be two boxes, one six inches, the other four inches deep without cover. Hinge them together so they will close as they were before being sawed in two. Near the top make three one-inch holes in the two ends for ventilation. For the hover make a frame of one-and-a-half by one-inch lumber, so it will fit inside the box. On the under side of this frame tack cloth loosely, so it will hang in the center nearly two inches below the frame. The cloth is to touch the chicks' backs. Nail cleats across the ends of the lower box to hold the frame in position. The top of the frame should be even with the top edge of the lower box. Cut a hole on the opposite side of the bottom box to the hinges, for the chickens to go in and out.

A friend who made this brooder tacked a piece of burlap on the floor and then filled it almost up to the cloth on the frame (the hover) with finely cut straw or hay. He then scooped out a nest in the center of it and put the baby chicks into it. The two-foot size is large enough to contain from one dozen to fifty chicks for one week, twenty-five till they are three weeks old, and twenty till they are six weeks old, or about that age. On very cold nights at first he put a little piece of blanket on top of the hover. As the chicks grew older he lessened the amount of straw or chaff, when the chicks were large enough to raise the heat sufficiently. After using this brooder (home made) all last winter, he said he would never be without it. Personally, I think it would be a good plan to let in a slide of glass at one side, as chickens do not like to go into a dark place. I do not know where you can get plans for making a brooder, but you can buy fireless brooders at any of the large poultry supply houses advertising in this book. This is Mr. Killifer's brooder.

———

DIPPING HENS—Would you be so kind as to write and let me know about dipping hens, etc? I have a flock of somewhere between five and six hundred. I notice some of them have lice and bunches of nits on their feathers. Whenever I have caught a hen I have greased her well, but this would take too long to go through the bunch. Is there any dip that would be strong enough and do no harm to the birds that would kill the nits with only one dipping?—W. B.

Answer—As you have so large a flock of hens and do not seem able or inclined to pull out the feathers that have nits on them, I think you will have to dip them twice, with an interval of five or six days. The nits are sure to hatch out in about five days after they are deposited by the lice, and by twice dipping them you should get most of them. It is an excellent plan in warm weather just at the commencement of the moult, to immerse the fowls in a diluted kerosene emulsion, wetting them thoroughly to the skin, or dip them in strong tobacco water, or a solution of two per cent creolin or chloro naptholeum. A well-known poultryman gives the following advice: Take the strongest and purest tobacco, 25 cents' worth being ample to clean off three hundred fowls. Make a decoction quite strong. If the user will observe a few points, no one will ever regret using tobacco to kill lice and not a solitary one will be left.

First, if the dipping is done out of doors, the thermometer should be at least 80 in the shade; second, the water should never be more than blood warm, say 98 degrees; third, and this is the most important point, every solitary feather must be made soaking wet, else you will not make a clean job of it. In dipping all fowls having heavy plumage, like the Brahmas and Cochins, the feathers must be raised with the hand and the water allowed to thoroughly wet the bird to the skin. This takes from one to two minutes for large, well-feathered fowls. If a dry feather is left there will be lice upon it. Do not dip the head under, but when the fowl is quiet, dip the head until all is under up to the eyes. When they will not hold still, use a small sponge and wet the top of their heads. No one who has fowls troubled with lice need fear to try this. It is very effective.

You must thoroughly clean the houses to get rid of the lice, and paint the perches with a good lice paint or liquid lice killer.

Give the hens a nice freshly dug up dust bath and they will keep themselves clean of lice. You can add one of the good lice powders to the dust bath if you wish.

FORMULA FOR CHICK FEED—The formula for chick feed that you want is as follows:

Chick feed for little chicks from the time they are hatched: 30 lbs. cracked wheat, 30 lbs. rolled or steel-cut oats, 15 lbs. finely cracked corn, 10 lbs. each of rich, millet, pearl barley, mustard or rape seed, granulated or ground bone, dried blood or granulated milk, chick grit, 5 lbs. granulated charcoal.

Mix and keep always before the chicks. Also clean water and skim milk if you have it. Note in the chick feed that wheat, oats and cracked corn are the chief ingredients. The others are to give a variety, and if you cannot get them, you just will have to leave them out. The bone and the dried blood are the animal part of the ration and can be substituted by fresh meat or milk or clabber or cottage cheese.

A formula for laying hens which I have used for years is: Two measures of bran, one measure of alfalfa meal, one measure of beef scraps, and in the breeding season one measure of oatmeal or rolled oats. This mixture can be used as a dry mash or mixed with water as a moist (but not sloppy) mash. I add a little pepper and salt to it to season it.

At moulting time I also add a quarter of a measure of linseed meal, or, if I cannot get that, half a measure of cottonseed meal, and sometimes a little tonic to help on the moult. The linseed meal gives a gloss to the new feathers that nothing else will give. The hens should have before them all the time good, sharp grit and oyster shells crushed. The oyster shells are to supply the lime to make the egg shell.

BROKEN DOWN HEN—There are two things I am anxious to know and I think you can help me from your experience. I have a hen whose hind part has been gradually swelling until now it nearly touches the ground. The feathers have all dropped out of her head. I think an egg may have been broken inside, but she seems so healthy that hardly seems possible. Please state cure, if any.— G. F. M.

Answer—Your hen has what we call a "break down." This is the result of a too fattening diet or too much corn, and too little of the muscle, bone-forming and egg elements. There is a large fat deposit in the abdomen, bulging and dragging down the skin and muscles, giving an ungainly appearance to the bird. It is a question whether to diet her or to eat her. I would advise the latter, as she will not prove a very good layer after this. The bareness of head also indicates an unbalanced ration and an insufficiency of "protein," the feather making element. A little carbolated vaseline rubbed in twice a week and more green food and more animal food in the ration will recify this.

FOR LAYERS—Will you please answer the following questions: Will hens lay as well without the male bird?

Which would you advise me to keep for breeders, pullets, hatched last spring, which are laying now, or the one-year-old hens?

Which is the best feed for them to produce eggs, the warm mash in the morning and corn at night or the dry feed—Mrs. O. G. L.

Answer—1. Yes, and the eggs will keep better.

2. Keep hens for mothers and pullets for your winter layers is the best rule.

3. I prefer to give the mash, if I give any, at night; then I can use up the table scraps, mixing them with bran, corn meal and alfalfa meal, giving the fowls either dry mash in hoppers or grain in their scratching pen, to induce them to exercise for their day meal. In this way I get more eggs.

TESTING OUT INFERTILE EGGS—I note in the paper an advertisement for an egg-tester which claims that it is possible to test out the infertile eggs before setting. Will you please tell me if you think this is possible? —Mrs. J. F. Y.

Answer — The advertisement which you mention was misleading. The way in which it tested the eggs was by floating them with the instrument in water; if they proved heavy enough to sink to a certain depth it showed that the egg was rich enough to support the life of a chick, should there be a germ in that egg. The machine could not show whether there was a germ in the egg, consequently it could not show if the egg was fertilized or not. The little germ is so infinitesimally small that it would make no appreciable difference in the weight of the egg.

PACKING EGGS FOR HATCHING—Will you kindly answer the following:
1. How long can one keep eggs for setting.

2. How is the best way to ship eggs for setting so they will not get broken? —Mrs. C. D. D.

Answer—1. You can keep your eggs three weeks or even more by turning them every day, but you must remember that the longer you keep them the fewer will hatch and they will not be as vigorous chicks as if the eggs had been fresh when set.

2. You can now get egg boxes made for packing eggs for expressing, or you can pack them in common slat baskets or peach baskets. I really prefer the baskets. I put a layer of excelsior in the bottom of the basket, then wrap each egg in a piece of newspaper about six inches square; set them little end down, packing excelsior between them, then put a layer of excelsior on the top, and cover with burlap, sewing it into the basket with twine. Mark plainly, "Eggs for hatching, handle with care." In the many thousands of eggs I have sent out, only two baskets had any broken eggs.

TURKEY QUESTIONS AND ANSWERS

TOMATOES FOR TURKEYS—I am feeding my turkeys a small ration of ripe tomatoes. Is this a proper food for them? —W. F. G.

Answer—A small amount of ripe tomatoes will not do your turkeys any harm. They are very fond of them, and it will benefit them, although there is very little nourishment in the tomatoes; the acidity seems to agree with them.

TURKEYS HAVE CHICKEN-POX—What is the matter with my young turkeys, and what shall I do for them? All over their heads and bills there are lumps forming like warts. Some of them have just a few, while others have their heads covered with them. The turkeys are about half grown. They are not penned up and have plenty of green alfalfa. We feed wheat and meat scraps occasionally.—Miss M. M.

Answer—Your turkeys have chicken-pox. The cure is to apply carbolic salve or carbolated vaseline. In three days bathe the affected parts with warm soapsuds in which are a few drops of carbolic acid, and again apply the salve.

Add a little sulphur to their food. This will hasten the cure. They should be cured in a little over a week. Be sure to separate all the fowls affected from the flock. This will prevent the spreading of the disease.

Dr. Haring of the University of California recommends to paint the spots or warts with iodine twice a week. This is rather a severe treatment but a sure cure.

TURKEYS LAME—Will you kindly tell me what to do for my turkeys? My early hatches did fine, but of the late hatch, four of them were troubled with stiff legs, one died, and one got well, but the other two are still lame, the knee joints are swollen and kind of pink color. Their appetites are good. —K. C.

Answer—Your turkeys have rheumatism. This comes from their liver being affected, by cold or damp weather. Give each of the affected turkeys a small liver pill, followed by a one-grain quinine pill every day for a week. Bathe the knee joints with the following: One cup of vinegar, one cup of turpentine, one heaping tablespoon of saltpeter. Mix, keep in a bottle, shake before using. I think this will cure them. Be careful not to give them any corn or corn meal, and give plenty of lettuce and onion.

GENERAL CARE OF TURKEYS—I would like to ask a few questions about turkeys. You mentioned raising them in a brooder. 1. How warm should one have the brooder when the poults are first put in? 1. At the end of the first week what should the temperature be lowered to? 3. Is alfalfa meal necessary or of any benefit to little poults or to little chicks if they have all the green barley they will eat, cut fine?—A Beginner.

Answer—The heat under the hover should be about 95. The reason I say "about" is that on a very warm, sunny day it might be a little lower, but should the outside temperature be cold or the weather damp and gloomy, it might be up to 95 for the best results. 2. About 85, depending somewhat on the outside air and weather. Gradually lower the temperature till you get it to 70 or 80, according to the weather. 3. No! Little turkeys require the succulent green, not the dried hay, ground up. Give them lettuce chopped up at first with every meal; then either lettuce, dandelion leaves, onion tops chopped fine, or cabbage or the tender leaves of beets. Any green vegetable that you would eat yourself will do and also the green barley as long as it is succulent and tender. Barley soon gets tough and hard and then it is not suitable for the little turkeys.

KEEP SEPARATE FROM CHICKS—Will you kindly give me some information concerning newly hatched turkeys? We have two hens and a tom. Would you advise keeping them away from chickens?—Mrs. C. B.

Answer—Little turkeys do much better when kept away from chickens. They require, or do better, on different food, and when very young require to be kept quiet, whilst the chicks like to scratch and rustle. Turkeys move more slowly and need rest and quiet. Then, again, corn, kaffir corn and corn meal suit chickens, but ferment inside the little turkeys and give them diarrhoea, which is often fatal. Let the turkey mothers take care of the little turkeys and give

them grass or alfalfa to run on and they will do well.

TURKEYS—I am glad if I have been able to help you with your turkeys, and will try to reply to your questions, but I wish you could give your turkeys free range as they are the Bronze, for that most beautiful breed is nearer to the wild than any other and, therefore, need more than any, a good wide free range to keep them healthy. A turkey on the range eats a few seeds, then sees an insect, maybe a grasshopper, and chases after that, which is good exercise. After a run he finds perhaps a nice little pebble or a few green leaves or twigs. and so on. He only eats a very little at a time and exercises between each mouthful and this is the way a turkey should eat. The nearer we can come to copying nature in feeding turkeys, the better success we shall have. Now, with this prelude I will try to answer your questions to the best of my ability. 1. How much grain and what kinds should I feed? 2. Should I give them bran and beef scraps? 3. Or do you prefer granulated milk? 4. How much of the milk should they have? 5. Should I feed more than twice a day? 6. Is there any food which should be always before them?—Mrs. C. F. S.

Keeping twenty young three-month-old turkeys yarded is a very serious proposition, unless your yard is an unusually large one with plenty of shade and sunshine. 1. Wheat is the best grain for turkeys until about two or three weeks before you want to kill them, then you can add corn. 2. You can give bran and beef scraps, but, 3, I prefer granulated milk and bran, as it seems to agree better with the turkeys. 4. About an ounce each per day. 5. Twice a day is considered about right for yarded turkeys. 6. Turkeys need plenty of fresh, green succulent food, such as clover, lawn clippings or lettuce, swiss chard, beet tops, cabbage or the curly kale. They must have green food to do well and should have all they can eat of it, and grain only twice a day. Almost any kind of fruit or nuts or olives suits them. If you want to leave any food always before them you might leave a box of granulated milk and another of bran. Always keep charcoal. grit and granulated bone before them. If you had a walnut orchard in which they could roam I would say leave a box of wheat where they can get to it and they will not overeat; they will roam away and only go to it when hungry, but in a yard with nothing to occupy or interest them, I think the bran would be better. Give them at least three or four times a week, onions chopped up and mixed with dry bran. The onions are a wonderful tonic to liver and kidneys and will do more to help you keep the turkeys healthy than anything. They are also a preventive to intestinal worms and roup. Fresh clean water as cool as possible is also a necessity.

A LACK OF GREEN FOOD—I have a tom turkey that is sick. He was a year old last May and about six weeks ago he would not eat. He did not look sick, and would strut and gobble a little, but did not eat. I gave him Carters' liver pills and he soon got all right. About a week ago he began to get off his feed again, and I at once began to doctor him. Have given him liver pills and germazone, but he has not eaten anything since last Wednesday. Can you tell me what ails him and what to do for him? He is a very valuable bird and I am anxious to have him get well. His usual feed is bran, barley meal, alfalfa meal and beef scrap in the morning and wheat and kaffir corn at night, with plenty of grit and oyster shell.—Mrs. G. H. B.

Answer—I think your turkey requires more green food than you are giving him, as you only mention alfalfa meal. Give him now, a quinine pill (two grains) every night for a week. Add charcoal and chopped onions to his mash in the morning, and plenty of green food once or twice a day. Give him as large a range as possible, or if you cannot give him range, let him out on your own lawn for two hours before sundown. What he needs is fresh green food and chopped onions for the liver tonic.

TURKEY'S CHICKEN-POX—I have some young turkeys several months old. On the heads of some are round things like warts; on one they are sore looking and are also on each knee joint of the legs. The turkeys don't appear sick. We have rubbed the heads with axle grease, as once before that seemed to help. What is the cause of this disease? How can one cure or prevent it and are the fowls good for food if they recover? My turkeys have free range and have plenty of animal food in the shape of

bugs, etc., all summer, also of course, green food in as large a quantity as they cared for. I have only fed them wheat. Chicken ticks, these flat bugs are bad here, but the turkeys roost outside, so should not be bothered much.—M. A.

Answer—Your turkeys have chickenpox. It comes from a microbe which gains entrance under the skin from some slight abraison, such as a scratch, or the bite of an insect. It is very prevalent during the fall, but except in the case of very young chickens, is easily curable, and the remedies you are using will effect a speedy cure.

Carbolic salve or Kileroup is the usual cure—or you can wash the spots in hot soapsuds to get off the scab and then grease just only the spots. The carbolic acid in the salve kills the microbe. The turkeys are perfectly fit for food. You had better be sure the ticks do not crawl up the trees to the turkeys. Pour a little stream of crude petroleum at the foot of the trees to keep off the ticks.

TURKEYS—Will you kindly tell how to raise little turkeys without any milk, or can't it be done? We value your writing very much.—H. D. C.

Answer—The milk that we use in feeding little turkeys, either as plain skim milk for them to drink or as a curd for them to eat, is given because it is found to be the best substitute for the insects that would be Nature's diet for the little turkeys. The next best substitute is hard boiled eggs, and after that ground-up meat, either raw or cooked.

Here in Los Angeles we can get the granulated and the dried milk, and these make good feed, both for turkeys and chickens. I should think you could get either of these at the poultry supply houses in Santa Cruz.

SICK GOBBLER—I write again in regard to a fine gobbler. He was hatched last May. He has been sick about ten days. Just sits around and does not walk much. Eats very little, and his droppings are nearly all white and small in quantity. His food has been rolled barley, wheat, and we have nine acres in green barley. He has plenty of clean, pure water and is not lousey, as I dust my turkeys with insecticide every week. When he first drooped around I gave him some liver pills, but he does not get much better. I hope you may be able to tell me something that will help him as I should feel very badly to lose him. —Mrs. S. H. J.

Answer—I would advise you first to stop dusting that gobbler with insect powder, as it may be disagreeing with him. Secondly, I would give him small liver pills, and at the same time, for at least a week, a pill of one or two grains of quinine every night. Also notice his droppings, if possible, because he may have intestinal worms, although the symptoms are more like kidney trouble.

TAPEWORM IN TURKEYS—I have over 100 turkeys that seem to be healthy but do not grow as they should. I find now they are full of long worms, probably tape worms. What shall I do?—Mrs. L. B. D.

Answer—If your turkeys have tapeworms, the best remedy I know is male fern (felix mas). It may be used in the form of a powder; (dose thirty grains to one dram) or of liquid extract (dose fifteen to thirty drops). It should be given in the morning and evening before feeding. Oil of turpentine is an excellent remedy for the common round worm; dose one to three teaspoonsful in an equal amount of castor oil. Feeding stewed garlic or raw onions will help the cure.

SHIPPING TURKEYS—Can turkey eggs be hatched successfully in an incubator or are they more apt to die? Will it hurt the little turkeys to be carried on the car any great distance?—Mrs. A. P.

Answer—Turkey eggs can be hatched in an incubator, if you don't mix them with other eggs, otherwise they do better under the hen. They can be raised in brooders, and it will not hurt them to travel on the cars if they do not get chilled.

How MANY TOMS?—I want to ask you how many turkey toms I should have for 24 hens. I have two fine toms weighing about 22 pounds each. Their beards are well developed and they appear to be very good birds. Will those two be enough for 24 hens?—Mrs. C. B. L.

Answer—It really would be better to have three toms, but under the circumstances I would rather risk having two good toms than to buy a third of unknown quality.

The rule is one yearling tom to ten hens. One tom will do for twenty hens sometimes, but ten hens is about the best number.

LIVER TROUBLE—We are in trouble with our little turkeys, and would like to ask you to help us. They were fine, strong fellows until a few days ago, when four of them suddenly died. I just noticed two of them, a little droopy in the afternoon, and four were dead the next morning. There was the slightest touch of diarrhoea noticeable, and I immediately put a little germazone in their water, and they have had it for several days. They have no signs of it now, but four more died last night, and several others are drooping. We made an examination this morning and found the liver all blotched and spotted all over in dark rings. That is all we could find wrong. The gizzard was healthy and full of grit and seemed perfect and in order.—Mrs. A. H.

Answer—The spotted liver is all that killed them. It denotes congestion of the liver. This is usually brought on by wrong feeding, or overfeeding, but it also comes from their taking cold; either from being too warm at night, under the chicken hen, getting them hot and sweaty, and then coming out in the morning into the cool, foggy air, which gives them a sudden chill. This would affect the liver, and make even the proper food disagree with them. They may take cold and get a chill affecting the liver, from running in damp alfalfa; or the chicken hen may drag them about and make the exercise too much, and this also would weaken their liver and make them susceptible to cold, which would affect their liver. I can only give you these suggestions, as I do not know all your conditions. One of the best remedies for diarrhoea in both chickens and little turkeys, is rice boiled in milk, with a tablespoonful of ground cinnamon to every pint of milk. Rice given even dry will help in a case of this kind.

ABOUT DUCKS

DUCK EGGS VS. HEN EGGS—What difference, if any, should there be in running an incubator with duck eggs from hen eggs? I am very successful with hen eggs but never succeeded very well with duck eggs; the same eggs hatch 90 per cent under a hen, and the first test from the incubator is about 90 per cent and then they die in the shell.—J. W. L.

Answer—Duck eggs require different treatment than the hen eggs. After the first test when you take them out to turn them, sprinkle them every day with warm water. Leave them out a few minutes to partially dry off, fan the stale air out of the incubator and then replace them. By this means I think you will have a better hatch. Duck eggs require more drying out than hen eggs and yet the shell must be dampened to make it brittle. Putting water into the incubator does not do as well as sprinkling.

FOOD—GOOD AND BAD—1. Would lettuce make good greens to sow in runways for Indian Runner ducks?
2. Will some whole wheat hurt them if they are provided with grit?
3. At what age should ducks hatched in March commence laying?
4. Will beef suet and chopped fresh beef do to feed them?—Mrs. F. H.

Answer—1. Lettuce is good for all fowls and would be good for the ducks as long as it lasts, but I am afraid the little fellows would soon pull it all up.
2. Whole wheat is not as good for little ducks as bran and corn meal. See article in this book.
3. Indian Runners hatched in March will commence laying in September.
4. Beef suet is not the food for ducks, but if you want to fatten them, you might add a little of it to their mash.

INDIGESTION—What is wrong with my ducks? They are almost full grown, and they turn over on their backs and are unable to get up; they are very weak; their eyes scale over and some of them have died. They act very much like chickens with the roup, only they do not swell around the head.—Mrs. J. G. C.

Answer—Your ducks are suffering from indigestion and also from their heads being stopped up. The indigestion comes partly from their not having sufficient sand with their food, and their heads being stopped up, comes from the drinking vessel not being deep enough so they can rinse their nostrils out many times during the day. If you remedy these two causes of trouble in the duck yard and feed them properly, giving but little whole grain, I think they will soon recover.

INCUBATOR DUCKS—We want to know the proper way to operate an incubator to hatch ducks. I have had fairly good luck hatching chickens but not with my ducks. I got only 40 out of 112 fertile eggs, and this time we should like to have a few directions to go by.
Do they require as much as chickens as to moisture; do you sprinkle, also how often, and as to airing the eggs, what time of day and how long do you advise to leave the machine open; how often do you test the eggs?—Mrs. W.

Answer—Duck eggs require quite as much heat as those of the chickens; they require more airing. Should be sprinkled with warm water once the first week, twice the second and every day thereafter, but do not put any water in the pans. Sprinkling the eggs helps to make the shells more brittle so the ducks will get out easier. Test the 5th day and again about once every week to take out the dead germs, as they putrify and are injurious to the rest. When you air the eggs, which you should do twice a day, that is every twelve hours, fan the stale air out of the incubator and then close up. Commence to air the eggs when you commence to turn them, that is 48 hours after they have been in the machine. The air space in the egg should be at the large end. I think if you follow the directions from the maker of the machine, and these hints, you will have a good hatch.

TO SECURE FERTILITY—I am starting to raise Indian Runner ducks and want to ask you how many ducks to put with one drake of this variety, so as to secure the highest possible fertility of eggs without keeping unnecessary drakes? I have a flock of 20 ducks and within a few days will be ready to start my incubator, so if you will kindly reply as soon as possible, I will be very much obliged to you.—L. F. R.

Answer—The number of Indian Runner ducks to one drake is ten. This has been found to be the best number for Indian Runners, although you can mate fifteen ducks to one drake and have good fertility. I want, however, to warn you that the eggs are not nearly so fertile in the fall and winter as they are in the spring, so you must not be disappointed if at least half of the eggs are infertile at this time of the year. To increase the fertility, would advise you to increase the amount of animal food you are feeding. You can tell in five days of incubation whether the eggs are fertile and those that are not fertile should be removed from the incubator and can be used for cooking or eating. They are merely infertile eggs that have been kept in a warm place for five days, and are better than most store eggs.

WEIGHT AT TEN WEEKS—Will you please inform me what weight most of the duck men can put on Indian Runner ducks at ten weeks?—I. L. R.

Answer—Indian Runners at ten weeks of age weigh as much as do the Pekins at that time, namely, about eight pounds per pair. They should be sent to market at from eight to ten weeks of age. After that the pin feathers develop, making them very hard to pick. I think you will be greatly pleased with the ducks when you try them. Their flesh is very delicious, fine grained and the bones are small. They have very much the flavor of the canvas-back, and I have heard, are sometimes sold instead of them. They are also the greatest layers of any known fowl; the eggs are white and very delicious, with no strong taste like the eggs of other varieties of ducks.

FEEDING FOR EGGS—I bought some Indian Runner ducks, thirty-six in all, and six drakes. They were laying up to the middle of December; since that time have layed none. I feed them about every thing that would come from a first-class hotel—bread, meat, oat and corn-meal mush, all kinds of vegetables and fruit. Three times a week I mix cracked corn and bran. I feed in the morning, twelve quarts, same amount at night. They have access to plenty of running water and keep perfectly clean. The pen is covered with forest leaves that makes it warm. What I want to know is, am I feeding right for laying later on? Is it customary to pick them? Does it affect their laying? I have over two hundred eggs engaged at 10 cents apiece. I want to raise all I can the comin season.—J. W. A.

Answer—I think that your hotel waste may have rather more bread in it than is good for egg production. Indian Runner ducks usually stop laying in October, commencing again in December, and getting into full lay in February. The best time for hatching Indian Runners is from the first of February to the end of July; the eggs are very fertile at such time. It may be that you are fattening the ducks too much, as overfat ducks do not lay well. They require much more animal food than chickens. In their wild state they live on grasses, fish, frogs and insects, with but very little grain. If you think they are getting too much bread, you might save some of it for chickens, and increase the amount of meat; keep them well supplied with coarse sand, grit and crushed oyster shells.

EGGS, GOOSE AND DUCK—I would like to know what care duck and geese eggs should have when a hen is sitting on them instead of the goose or duck. Also, what feed should they have when first hatched?—Mrs. J. A. P.

Answer—Goose and duck eggs require more heat and a longer period of incubation than hens' eggs. Five goose eggs are sufficient to place under a hen, and be sure that she turns the eggs every day or the gosling will be a cripple. The goose eggs are heavy for a hen to turn, and for this reason, and also because they require more heat, the hen should not have more than five to care for. From nine to eleven duck eggs are the number, for the same reasons, that should be given to a hen.

Goose eggs require thirty days of incubation; duck eggs twenty-eight. Hens are apt to desert them toward the last and should be watched, as they get tired of waiting for their chicks to come out. I also have had hens that were so much afraid of the queer, green looking babies they hatched out that they would kill them. They seem to know that they are not proper chickens. I feed the little geese hard-boiled eggs, chopped fine, and cracker crumbs moistened with water, and sprinkle a little sand on the food. This is the first food. The next day they get the same, with lettuce chopped fine. After this I add breakfast oats with it and bran. As early as possible I put the geese out on the lawn, take the hen away from them and put them into a box in the woodshed or kitchen, if the nights are cool, or if I am afraid of cats or other marauders. They do not require heat after a few days, sometimes not after the first day. It depends upon the weather.

Geese are the easiest of fowls to raise. They are a grazing bird and must have a pasture of something green to graze on. When young, they should not have whole grain, but a mash of bran and corn meal with a little animal food in it, and always grass or alfalfa to graze on.

Ducks do well treated in the same way, remembering to give them a little sand with each meal.

——

Died in the Shell.—I had two hens sitting on duck eggs and the ducks all died in the shell. The eggs were pipped, but it seemed as though the ducks could not get out. I dipped the eggs the last six days in lukewarm water once a day. I opened two eggs and there was jelly around the ducks. Could you kindly let me know why and how it is, as I have two more hens setting?—Mrs. C. F. N.

Answer—Sprinkle your duck eggs, if the weather is warm and dry, three times a week after the first week; let the water be just as hot as you can bear your hand in, and sprinkle it out of a little sprinkling pot or use a whist broom to sprinkle the eggs with as you would clothes for ironing; leave the eggs damp for the hen to go on them. This is better than floating them in the water. Little ducks can be easily helped out of the eggs and still live and be strong; if they seem slow in hatching, bring them into the house and put a warm damp flannel around them and place at the back of the kitchen stove, and I think they will then come out without assistance; if not, help them out.

——

GEESE

Geese—I have a few geese and just lately they have started to lay; gather from four to six daily. Do you think by turning them daily I might save them up for incubation? About what degree should be kept up for them? I put seven eggs under a hen. Would you also tell me what should baby geese be fed?—J. W.

Answer—You can keep geese eggs, by turning them every day, for three weeks. They take thirty days to incubate. The incubator should be about 102½ for the first week and 103 afterwards. Five eggs is plenty to put under a hen. See instructions in this book for hatching duck eggs in an incubator. Treat goose eggs in the same way. Feed baby geese the same as baby ducks for the first week, gradually adding chopped lettuce until at least half their food is green food. Geese are grazing animals and require plenty of green, succulent food. They are very easy to raise and do not require brooder heat more than a few days.

——

Toulouse Geese—First, I have a few geese. I had eight Toulouse goslings. I fed them boiled eggs, bread crumbs, oatmeal (dry), and sometimes clabber cheese with a lot of fine cut grass and young rye from the rye patch, as I have no lettuce yet, plenty of gravel and a pan of water, but they all die from a week to three weeks of age. Now, what is the cause and what can I do to raise the others, as I hate to lose them so bad.—Mrs. J. B. M.

Answer—You feed your young geese wrong. Geese are grazing animals and need grass or young tender clover to eat. Next time you have any give them bran (three cups full) and corn meal (one cup full) moistened with water, with a teaspoonful of sand sprinkled over it. This should be fed every two hours, after the first day, when they should have nothing at all to eat, they should be turned out on the grass or on a clover lawn. From the very first they must have grass or clover to crop from. After the first week leave the food where they can get it all the time and they will feed themselves without any trouble. Geese are the easiest of all fowls to raise. They must not have water to swim in until they have their mature feathers.

White Holland Turkeys
THE BEST

LITTLE MONEY-MAKERS

from stock taking first, second and third prizes at Los Angeles Show, 1912, at *eight months weigh 20 pounds.*

Our strain of these turkeys is beautifully white, vigorous, hardy, large-sized, early maturing, easily adapted to large yards. They are home stayers, splendid layers, and an excellent market fowl.

We can supply eggs for hatching and high-class breeding stock at reasonable prices. Address

MRS. W. D. ROOT,

Box 105, R. D. 1, Los Angeles, California.
Residence, 735 Verdugo Road, Glendale.

Single Comb Black Minorcas
are the
BIG BIRDS THAT LAY BIG WHITE EGGS

and keep at it the year around. A splendid general purpose fowl, they PAY THEIR WAY where others fail.

In four years, 1908-1911, of prize winning at Los Angeles shows, our birds took 4 firsts, 2 seconds, 2 thirds, 1 fourth, 1 fifth, proving conclusively that ours is first-class stock, equal to the best, up-to-date in weight, color and shape.

Eggs from carefully mated pens—only $2.50 per 15; $4.00 for 30. Stock a matter of correspondence.

MRS. W. D. ROOT,

Address, Box 105 R. D. 1, Los Angeles, California.
Residence, 735 Verdugo Road, Glendale, Cal.

(Take Glendale Pacific Electric at 6th and Main, Los Angeles, get off at 4th St., Glendale, take Eagle Rock car (5 min.) to Verdugo Road, get off and walk south 4 blocks.)

Red
Feather
Farm

ORPINGTONS

BLACK Prize winners.
Our foundation stock came direct from England—

BUFF Blue ribbon winners in California, imported direct from England.

Pure Penciled Fawn and White "Wonder" Indian Runners

from the celebrated stock of S. H. Scott of New Zealand. Pedigreed for four generations of the greatest egg layers in the world. Parents of these "Wonder" ducks laid 308 to 320 eggs per year.

Great winter layers; eggs white and large.   Catalogue free. Stock and eggs for sale.

MRS. M. E. PLAW

2620 East 27th St., Fruitvale, Cal.

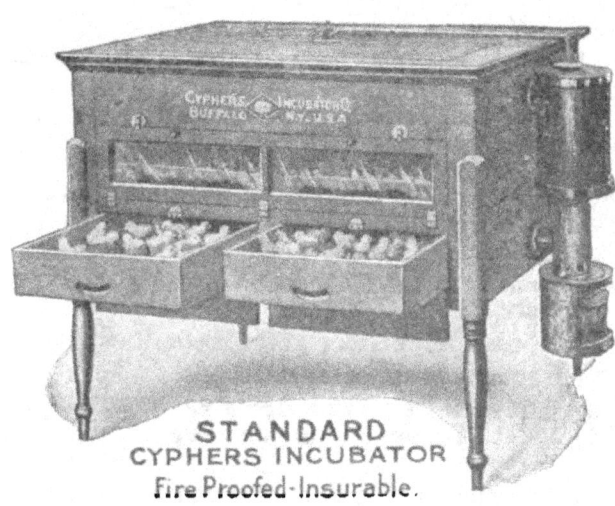